SpringerBriefs in Physics

SpringerBriefs in Physics are a series of slim high-quality publications encompassing the entire spectrum of physics. Manuscripts for SpringerBriefs in Physics will be evaluated by Springer and by members of the Editorial Board. Proposals and other communication should be sent to your Publishing Editors at Springer.

Featuring compact volumes of 50 to 125 pages (approximately 20,000–45,000 words), Briefs are shorter than a conventional book but longer than a journal article. Thus, Briefs serve as timely, concise tools for students, researchers, and professionals. Typical texts for publication might include:

- A snapshot review of the current state of a hot or emerging field
- A concise introduction to core concepts that students must understand in order to make independent contributions
- An extended research report giving more details and discussion than is possible in a conventional journal article
- A manual describing underlying principles and best practices for an experimental technique
- An essay exploring new ideas within physics, related philosophical issues, or broader topics such as science and society

Briefs allow authors to present their ideas and readers to absorb them with minimal time investment. Briefs will be published as part of Springer's eBook collection, with millions of users worldwide. In addition, they will be available, just like other books, for individual print and electronic purchase. Briefs are characterized by fast, global electronic dissemination, straightforward publishing agreements, easy-to-use manuscript preparation and formatting guidelines, and expedited production schedules. We aim for publication 8–12 weeks after acceptance.

More information about this series at http://www.springer.com/series/8902

Vidya Sagar Bhasin · Indranil Mazumdar

Few Body Dynamics, Efimov Effect and Halo Nuclei

 Springer

Vidya Sagar Bhasin
Department of Physics and Astrophysics
University of Delhi
New Delhi, India

Indranil Mazumdar
Department of Nuclear and Atomic Physics
Tata Institute of Fundamental Research
Mumbai, India

ISSN 2191-5423 ISSN 2191-5431 (electronic)
SpringerBriefs in Physics
ISBN 978-3-030-56170-3 ISBN 978-3-030-56171-0 (eBook)
https://doi.org/10.1007/978-3-030-56171-0

This Springer imprint is published by the registered company Springer Nature Switzerland AG
The registered company address is: Gewerbestrasse 11, 6330 Cham, Switzerland

Preface

The subject of few body physics has undergone a long journey during the past few decades, witnessing an upsurge of research activities since the pioneering work of Faddeev in 1960. While these activities are well documented and published in standard journals and review articles, we here wish to present only a brief historical perspective which led to this upsurge [1–6].

Faddeev [7], starting from the Lippmann–Schwinger equation for the three-body scattering problem, showed how the problem of disconnectedness of the scattering matrix appearing in the form of δ-type singularities inherently present in the Lippmann–Schwinger equation can be resolved leading to the three-body kernel which is completely continuous. Subsequently, Lovelace in 1964 [8], following Faddeev equations, undertook a detailed study and found that in situations where the two-body systems involved have bound or resonant states, the two-body T-matrices could be represented in a separable form with the result that resulting integral equations for the three-body scattering amplitude would involve only a single variable. Such equations could be easily handled computationally without resort to further approximations. Prior to this, Mitra in 1960 [9] had already initiated a three-body approach using two-body separable potentials and in 1963, Mitra and Bhasin [10] had derived, using three-body Schrodinger equation, the same equations as obtained by Lovelace for the n-d scattering problem. Independently, Sitenko and Kharchenko [11] also obtained similar equations using Faddeev's theory and employing separable two-body T-matrices. During the same period, Amado and his group [12], starting from the Lee model, developed a three-body model for n-d scattering which eventually led to similar equations obtained with separable potentials. Thus, these attempts carried out altogether during 1960–65 sparked off a spectacular interest in research activities in few body nuclear, atomic and molecular systems at various places all over the world. The major impact of this was the decision to hold regular international conferences after a period of every two years in different parts of the world to sustain the interest and encourage the young researchers in this field of few body physics.

Efimov in 1970 [13] discovered a remarkable phenomenon that can occur in three-body sector for non-relativistic particles when at least two of the three pairs of particles have a large scattering length. Using the hyperspherical approach for the three-body system, he pointed out that when the two-body scattering length $|a|$ is sufficiently large compared to the range r_0 of the potential, there is a sequence of three-body bound states whose binding energies are spaced roughly geometrically in the interval between $\hbar^2/(mr_0^2)$ and $\hbar^2/(ma^2)$. As a is increased, new bound states appear and, in the limit, as $a \to \pm\infty$, there are infinitely many three-body bound states accumulating near the three-body scattering threshold. An important feature of the Efimov effect is that the sequence of three-body bound states has universal properties which are insensitive to the details of the two-body potential at short distances. Dating back to1970 up to the recent times, tremendous progress in research of universal Efimov physics in the field of nuclear, atomic and molecular physics has taken off, driven particularly by a combination of experimental and theoretical studies [14].

With the advent of facilities for producing intense radioactive nuclear beams through high-energy fragmentation along with the developments in the state of the art involving detectors and isotope separators, experiments were carried out during 1990s at various laboratories which opened a new rich field of nuclear physics—the neutron-rich unstable nuclei. Such light exotic nuclei lie almost on the limit of nuclear stability corresponding roughly to zero binding energy of the last few neutrons and are commonly referred as neutron drip line nuclei. The pioneering experiments [15] which were carried out to investigate the properties of these so-called halo nuclei such as ^8He, ^{11}Li, ^{11}Be, ^{14}Be, ^{17}B, ^{20}C and ^{22}C, etc., revealed several striking features such as (i) abnormally large interaction cross sections, (ii) very weak binding of the last few nucleons leading to systems with qualitatively new structures and surface densities and (iii) large dissociation cross sections of a halo nucleus by a high Z-target of a new exotic mode of collective vibrations in the nucleus.

From the theoretical standpoint, these experimental findings posed serious challenging problems in studying the structural properties of such halo nuclei using conventional models, as nuclear shell model and Hartree–Fock formalism [16]. In certain cases, there was sufficient experimental evidence for the existence of a structure-less core with two loosely bound neutrons orbiting around it. A typical example is the ^{11}Li nucleus where experimental measurements [17] on the magnetic dipole and electric quadrupole moments give values which are close to those obtained for ^9Li. Such halo nuclei are the natural candidates for studying their ground state properties within the framework of three-body formalism. A characteristic property observed in such nuclei is that for a two-body system such as $n - {}^9Li$; there exists no bound state but as soon as we bring a second neutron, it forms a loosely bound three-body system. For this reason, such nuclei were named as 'Borromean nuclei' ('Borromean' refers to the famous three-ring system inter-linked in such a way that breaking-off any pair causes all the three rings to separate).

We had undertaken this project from early nineties to investigate the structural properties of such halo nuclei. Soon after we realized that these nuclei with halo structure and very weak binding characterized by large n-n scattering length could well be suited to study the Efimov effect. Subsequently, in 2n-halo nuclei like $^{14}Be, ^{19}B, ^{22}C$ and ^{20}C, we carried out the analysis of the occurrence of Efimov states. The analysis showed that Borromean-type nuclei, where n-n and n-core are both unbound, are much less vulnerable to respond to the existence of Efimov states. On the contrary, nuclei like ^{20}C in which the halo neutron is supposed to be in the intruder low-lying bound state with the core appear to be promising candidates for the occurrence of the Efimov states at energies below the n-(nc) breakup threshold. This motivated us to extend our study of the scattering of neutron on a bound $(n - {}^{18}C)$ system and showed that the Efimov states in ^{20}C move over to the physical scattering region causing a resonance in $n - {}^{19}C$ scattering around neutron incident energy of 1.6 keV. The scope of this work was enlarged beyond ^{20}C to include heavier nuclei like ^{32}Ne and ^{38}Mg in which the two neutrons are very weakly bound to the core.

In addition to the study of Efimov effect in halo nuclei, the structural properties of ^{11}Li as for instance (i) matter radius, momentum distributions of the halo neutrons and the core and n-n and n-c correlations, (ii) the resonant states of ^{11}Li in the continuum and (iii) β-decay of ^{11}Li were also investigated. The details of these results along with the discussions will be presented in the following chapters.

Another area in the field of universal Efimov physics which has witnessed tremendous progress during the last decade is the subject of ultra-cold atomic systems. The first experimental evidence of Efimov resonance was reported in 2006 by Kraemer et al. [18] in an optically trapped ultra-cold gas of cesium atoms. Experimentally, its signature as a giant three-body recombination loss was noticed by magnetically tuning the s-wave scattering length in a range between $-2500\ a_0$ and $+1600\ a_0$ (a_0 being the Bohr radius). In addition to the prominent three-body Efimov resonance at $a_1^{3b} = -850a_0$, a broad, weaker resonance feature was visible at a scattering length near a = -370 Bohr radii. The agreement of its position with the predicted universal four-body resonance location suggests that this feature is in fact associated with four-body loss and represents the lower energy tetramer resonance state. The four-body nature of those loss features was subsequently verified in 2009 by Ferlaino et al. [19]. Before these universal features were observed experimentally, a number of independent theoretical studies, carried out during 1999 and 2006 [20], provided significant quantitative understanding of the link between three-body recombination and universal physics. As a result, it appears that many body gas under suitable conditions has a promise to display a tremendously rich variety of phases in the universality range of large negative scattering lengths. Indeed, at a theoretical level, quantitative calculations of the four-body recombination loss rate in the universal regime have recently been carried out as an application of more general developments in the theory of N-body recombination [21].

The subject of effective field theory has, in recent years, attracted a great deal of attention to provide a powerful method for describing non-relativistic quantum mechanical systems composed of particles with wave numbers k much smaller than the inverse of the characteristic range R of the effective potential. The underlying premises of this approach are that particles with short-range interactions and a large scattering length have universal low-energy properties that do not depend on the details of their interactions at short distances. Effective field theory has recently been used to investigate the properties of ultracold atomic systems including the calculations of three-body recombination into deep bound states in Bose–Einstein condensates, the Efimov effect in halo nuclei and in general studying the low-energy properties of three-body scattering systems. Braaten and Hammer [22] have carried out detailed and systematic studies to investigate the universal properties of few body systems in atomic and nuclear physics within the framework of effective field theories introducing the concepts of renormalization group.

This brief introduction to the historical development of theoretical techniques of few body physics, the discovery of the Efimov effect and their application to understand the amazing structural features of the neutron-rich halo nuclei summarize the intent and scope of this book. Our efforts have been to provide an overview of the few body scattering theory and techniques developed in nuclear physics and their applications to explore the structural properties of neutron-rich unstable nuclei, primarily, the so-called halo nuclei. Readers will gain in-depth knowledge about the methods involved to solve the two- and three-body scattering problems and a special focus is put on the Faddeev approach and the power of separable potentials.

This book makes no attempts or is expected to replace classical and more formidable texts on scattering theory, namely by Roger Newton, or Watson & Goldeberger, or Charles Joachain. We basically provide the essentials of formal scattering theory of two- and three-body systems so as to prepare the reader for their practical application in the study of Halo nuclei and the Efimov effect in the following chapters. Formal theory of two- and three-body scattering has been discussed in Chaps. 1 and 2 in a compact and abridged form to initiate the beginners who want to attack the problems of halo nuclei within the framework of three-body models. In this sense, we address both the graduate students and senior researchers. Chapter 3 is devoted to a detailed analysis of the Efimov effect in three-body systems. We introduce the basic underlying physics of the Efimov effect in this chapter and subsequently investigate several two-neutron halo nuclei, both Borromean and non-Borromean to find whether they are the right candidates to support Efimov state(s) in Chap. 6. We also discuss in detail, drawing primarily from our work, how to analyze, within the framework of a three-body approach and using realistic short-range forces, the structural properties of halo nuclei, namely binding energies, momentum distributions, neutron-neutron and neutron-core correlations in two-neutron halo nuclei and the very challenging problem of beta decay of two-neutron halo nuclei in Chap. 6. In Chap. 4, we discuss the recent progress related to the use of effective field theories for investigating the properties of halo nuclei. We derive the three-body scattering amplitude in Effective Field Theory approaching from separable potentials and also present the computational details for the elastic scattering of neutrons from the loosely bound ^{19}C ($n - {}^{18}C$ system). Chapter 5 aims to provide, in a nutshell, the historical and global developments in experimental research in the field of halo nuclei and the most salient structural features of neutron-rich halo nuclei which have emerged from the Radioactive Ion Beam (RIB) facilities. The basic intention has been to provide the young and uninitiated researcher the proper perspective before embarking upon more focused theoretical work. We have primarily concentrated on the structural aspects of halo nuclei. A vast and emerging field of nuclear reactions with loosely bound halo nuclei has not been touched upon. However, we have briefly mentioned this field of research in Chap. 5 and have provided useful references by leading practitioners in the field. In spite of the best efforts of the authors, a technical monograph of this nature is unlikely to be totally unblemished of errors, technical or typographical. We would appreciate the attentive reader bringing to our notice any shortcoming of this nature.

New Delhi, India
Mumbai, India
June 2020

Vidya Sagar Bhasin
Indranil Mazumdar

Acknowledgements

The proposal to write this monograph came from Prof. B. Ananthanarayan, Editor of Springer Briefs in Physics series. Writing such a monograph was always dormant within us and the invitation was primarily responsible for bringing a long-cherished desire to fruition. Anant has been an exceptional editor, encouraging and ever ready to extend the deadline showing remarkable patience. We are pleasurably indebted to him. Thanks also go to Lisa Scalone of Springer for her patience and providing all guidance during the course of its writing. Chapters 4 and 6 draw heavily from our long-standing collaboration with several colleagues to understand the halo nuclei and the Efimov effect. We gratefully acknowledge contributions of Dr. V. Arora, Dr. S. K. Chauhan and Prof. A. Ravi P. Rau. We were initiated in the world of Fano resonances by ARPR with his encyclopedic knowledge of the subject. We express our sincere thanks to him. We thank Mr. V. Ranga for his timely help with the figures and overall adjustments of the text. We are also grateful to Mr. R. Sariyal for his help in adjustments of the text.

Contents

About the Authors

Vidya Sagar Bhasin received his Ph.D in theoretical nuclear physics from University of Delhi, working on the problem of nuclear three body scattering at low energies. He has spent extended periods at ICTP, Trieste, SLAC, Stanford University, T.W. Bonner Nuclear Lab, Rice University, University of Manitoba, Winnipeg at different times as post-doctoral fellow and visiting scientist. He has been Alexander von Humboldt Fellow at the University of Bonn. He taught and carried out research at the Department of Physics and Astrophysics, University of Delhi for nearly four decades. His research interests have been in non-relativistic and relativistic three-body nuclear physic, high energy scattering problems, meson-baryon physics in quark model, charge symmetry breaking effect in nucleon-nucleon scattering, few-body analysis of structural properties of neutron-rich halo nuclei and search for Efimov effect in nuclei. He has edited several volumes on some of these topics and has authored several books on science and technology for the National Centre for Education Research and Training (NCERT, Govt. of India).

Indranil Mazumdar is professor of nuclear physics at the Tata Institute of Fundamental Research (TIFR), Mumbai. He read physics at Delhi University obtaining both his bachelor and master degrees. He received his PhD in nuclear physics from Delhi University having worked jointly at Delhi University and Inter University Accelerator Centre (Formerly Nuclear Science Centre), New Delhi. He did his post-doctoral work at the State University of New York at Stony Brook, and later spent extended period on sabbatical at the Duke University, North Carolina. His research interests cover both theoretical calculations, experiments and instrumentation. His primary activities include few-body calculations of neutron-rich halo nuclei, Efimov effect, shape-phase transitions in hot and rotating nuclei, giant dipole resonances, fusion-fission dynamics, nuclear level density, reactions of nuclear astrophysics and Big Bang Nucleosynthesis.

Chapter 1
Essentials of Non-relativistic Two-Body Scattering Theory

1.1 Introduction

In this section, we retrace the basic steps required to formulate scattering theory in non-relativistic quantum mechanics with two-body potential [23]. We start with a time-dependent Schrodinger equation for the scattering state which is to be developed from a wave packet that is free in the infinite past. The requirement that the scattered wave packet coincides with the free packet demands that the norm of the difference between the two be zero in the strong limit. This leads us to introduce and define Moller operator. Although the domain of Moller operator is the Hilbert space of square-integrable free states, the work by Faddeev shows that this domain can be extended to include states of sharp momentum. This helps us to replace the strong limit by an Euler limit. With the introduction of the Euler limit, we are able to get two-particle scattering theory which is time independent.

We then introduce Green's function or the resolvent through which the scattering state and the free particle states can be related. The next step is to derive the integral equation for the scattering states incorporating proper boundary condition which consists of an incident plane wave plus an outgoing scattered wave. Next, we examine whether the kernel of the integral equation so obtained is a compact operator, or, in simple terms, whether the Schmidt norm is finite or not. We find that Schmidt norm can only be finite if (i) the integral over the absolute square of the potential exists thereby ruling out the possibility of long infinite range potentials and (ii) the imaginary part of $\sqrt{2\mu(E + i\varepsilon)}$ is zero. The latter condition excludes the scattering energies in which we are interested. However, a finite Schmidt norm is only a sufficient but not the necessary condition. Lovelace in his paper shows that the kernel of the integral equation is compact in the Banach space of continuous bounded functions with continuous derivatives. This ensures that the Lippmann–Schwinger equation has a unique solution.

To get the link between scattering states and the experimentally measured data, we introduce and define the S-matrix and obtain its relationship with the T-matrix

V. S. Bhasin and I. Mazumdar, *Few Body Dynamics, Efimov Effect and Halo Nuclei*, SpringerBriefs in Physics, https://doi.org/10.1007/978-3-030-56171-0_1

which is found to satisfy inhomogeneous integral equation. Finally, for a Hermitian potential, we have the unitarity relation for the S-matrix which ensures conservation of probability flux. This leads us to establish the well-known optical theorem relating the imaginary part of the forward amplitude to the total cross section.

1.2 Boundary Condition of the Scattering State: Basic Concepts

The scattering of two particles is described by the time-dependent Schrodinger equation:

$$i\hbar\frac{\partial \psi_a^{(+)}(t)}{\partial t} = H\psi_a^{(+)}(t), \tag{1.1}$$

where H is the Hamiltonian which consists of (i) the kinetic energy, $p^2/(2\mu) = H_0$, of the relative motion of the two particles plus (ii) the potential V, which is a function of the relative coordinates of the particles. The scattering state $\psi_a^{(+)}$ develops from a wave packet which is free in the infinite past and whose properties are characterized by the index a. The solution of Eq. (1.1) is given by

$$\psi_a^{(+)}(t) = \exp(-iHt/\hbar)\ \psi_a^{(+)} \tag{1.2}$$

The boundary condition is formulated by introducing a reference wave packet ϕ_a at $t = 0$ and its time dependence is given by the free Hamiltonian H_0:

$$\phi_a(t) = \exp(-i\ H_0 t/\hbar)\phi_a \tag{1.3}$$

It is required that in the infinite past the wave function $\psi_a^{(+)}$ coincides with the free packet ϕ_a. In the limit of $t \to -\infty$, since the wave packets spread out in time, it is not advisable to demand a pointwise agreement. That is why the strong limit is needed [Note, however, in real experiment the wave packets associated with the beam and detector are highly localized over experimental time scales.]. Thus, instead of requiring a pointwise agreement, we demand that the norm of the difference be zero, i.e.,

$$\lim_{t\to\infty}\left\|\exp(-iHt/\hbar)\psi_a^{(+)} - \exp(-iH_0t/\hbar)\phi_a\right\| = 0 \tag{1.4}$$

This limit on the norm, which is called a strong limit, is denoted by 's-lim,' which implies a certain restriction on the range of the potential. Equation (1.4) can thus be expressed as:

$$\psi_a^{(+)} = s-\lim_{t\to-\infty}\exp(iHt/\hbar)\exp(-iH_0t/\hbar)\phi_a \equiv \Omega^{(+)}\phi_a \tag{1.5}$$

1.3 Moller Operator

The right-hand side of Eq. (1.5) introduces a symbol $\Omega^{(+)}$, which operating on a free state produces the scattering state $\psi_a^{(+)}$. This defines the Moller operator $\Omega^{(+)}$, which is explicitly written as:

$$\Omega^{(\pm)} = s - \lim_{t \to \mp\infty} \exp(iHt/\hbar)\exp(-iH_0 t) \tag{1.6}$$

Clearly, the Moller operator $\Omega^{(-)}$ operating on the free state produces the scattering state $\psi_a^{(-)}(t)$, which approaches the free state $\phi_a(t)$ for $t \to +\infty$. From the definition (1.6), it follows that for any finite τ

$$s - \lim_{t \to \mp\infty} \|\exp(iH(t + \tau/\hbar)\exp(-iH_0(t + \tau)/\hbar)\|$$
$$= \Omega^{(\pm)}(\tau) = \exp(iH\tau/\hbar)\Omega^{(\pm)}\exp(-iH\tau/\hbar) \tag{1.7}$$

is also valid. By differentiating with respect to τ, we get

$$\frac{d\Omega^{(\pm)}}{d\tau} = \exp(iH\tau)\big(H\Omega^{(\pm)} - \Omega^{(\pm)}H_0\big)\exp(-iH_0\tau) = 0,$$

which gives the commutation relation

$$H\,\Omega^{(\pm)} = \Omega^{(\pm)}H_0 \tag{1.8}$$

(Note from here onwards we are using the units for which $\hbar = 1$).

The domain of Moller operator is the Hilbert space of square-integrable free states. However, work by Faddeev [7] has shown that this domain can be extended to include states of sharp momentum. We are thus allowed to use momentum eigenstates $|\vec{p}\rangle$ instead of the wave packets ϕ_a and apply the operators $\Omega^{(\pm)}$ on them to get scattering states $|\vec{p}\rangle^{(\pm)}$. Thus, for instance, the operator $\Omega^{(\pm)}$ operating on the state $|\vec{p}\rangle$ gives the scattering state $|\vec{p}\rangle^{(+)}$ consisting of a plane wave $\exp(ikz)$ which describes the incoming flux and the flux of the unscattered particles plus a spherical wave $f(\theta, \varphi)\exp(ikr)/r$ which describes the flux of the scattered particles. The scattering states $|\vec{p}\rangle^{(\pm)}$ have a sharp energy E just as the free states $|\vec{p}\rangle$ have the sharp energy $p^2/2\mu$. The two energies are the same, as can be shown by using the commutation relation (1.8):

$$H|\vec{p}\rangle^{(\pm)} = H\Omega^{(\pm)}|\vec{p}\rangle = \Omega^{(\pm)}H_0|\vec{p}\rangle$$
$$= \Omega^{(\pm)}\frac{p^2}{2\mu}|\vec{p}\rangle = \frac{p^2}{2\mu}|\vec{p}\rangle^{(\pm)} \tag{1.9}$$

However, applying the Hermitian conjugate operator $\Omega^{(\pm)*}$ on the bound states ψ_n of H, we find

$$\langle \Omega^{(\pm)^{*}} \psi_n \mid \vec{p} \rangle = \langle \psi_n | \Omega^{(\pm)} | \vec{p} \rangle = \langle \psi_n \mid \vec{p} \rangle^{(\pm)} = 0 \qquad (1.10)$$

because the states ψ_n and $| \vec{p} \rangle^{(\pm)}$ are orthogonal belonging to different energies. The time limit in Eq. (1.6) can be replaced by Euler's limit

$$\Omega^{(\pm)} = s - \lim_{t \to \mp \infty} \exp(i H t) \exp(-i H_0 t)$$

$$= s - \lim_{\varepsilon \to 0} \pm \varepsilon \int_{\mp\infty}^{0} \exp(\pm \varepsilon t) \exp(i H t) \exp(-i H_0 t) \mathrm{d}t \qquad (1.11)$$

Note that:

$$\lim_{\varepsilon \to 0} \int_{-\infty}^{0} \mathrm{d}t \varepsilon \exp(\varepsilon t) f(t) = f(\infty) \int_{-\infty}^{0} \mathrm{d}x \exp(x) = f(\infty) \qquad (1.12)$$

By introducing the Euler limit, the time dependence in the scattering theory can be eliminated. We shall see in the next section that following this procedure, a time-independent integral equation which incorporates the boundary condition can be obtained.

1.4 Resolvent Equation and Lippmann–Schwinger Equation

We can not carry out the integration in the expression (1.11) for the Moller operator for the simple reason H_0 and H do not commute with each other, i.e.,

$$\exp(i H t) \exp(-i H_0 t) \neq \exp(i (H - H_0)t)$$

But since one can apply the Moller operators on the plane states, we can write

$$|p\rangle^{(\pm)} = \lim_{\varepsilon \to 0} \pm \varepsilon \int_{\mp\infty}^{0} \mathrm{d}t \exp(\pm \varepsilon t) \exp(i H t) \exp(-i E t) | \vec{p} \rangle$$

$$= \lim_{\varepsilon \to 0} \pm \varepsilon \int_{\mp}^{0} \mathrm{d}t \exp[i (H - E \mp i\varepsilon)t] | \vec{p} \rangle$$

$$= \lim_{\varepsilon \to 0} \pm i\varepsilon [E \pm i\varepsilon - H]^{-1} | \vec{p} \rangle \qquad (1.13)$$

This enables us to introduce the definition of resolvent or Green's function:

$$g(z) = [z - H]^{-1}, \quad \text{where } z = E \pm i\varepsilon \tag{1.14}$$

which operating on the plane wave state $|\vec{p}\rangle$ gives the scattering states $|p\rangle^{(\pm)}$, viz.

$$|p\rangle^{(\pm)} = \lim_{\varepsilon \to 0}\left[\pm i\varepsilon g(E \pm i\varepsilon)|\vec{p}\rangle\right] \tag{1.15}$$

This is how we arrive at a time-independent theory starting from time-dependent approach.

Now using the relation (1.14), we write

$$z - z' = g^{-1}(z) - g^{-1}(z')$$

and multiplying both sides by $g(z)\,g(z')$ we get

$$g(z') - g(z) = (z - z')\,g(z)\,g(z') \tag{1.16}$$

With the definition of the resolvent for a free particle, i.e.,

$$g_0(z) = (z - H_0)^{-1} \tag{1.17}$$

we write

$$V = H - H_0 = g_0^{-1}(z) - g^{-1}(z)$$

On multiplication with $g(z)\,g_0(z)$, we get the resolvent equation:

$$\begin{aligned} g(z) &= g_0(z) + g_0(z)V\,g(z), \\ &= g_0(z) + g(z)V\,g_0(z) \end{aligned} \tag{1.18a, b}$$

Let us now derive, using this operator identity, the integral equation for the scattering states. Thus, from Eq. (1.15), we have

$$\begin{aligned} |p\rangle^{(\pm)} &= \lim_{\varepsilon \to 0}\left[\pm i\varepsilon\,g(E \pm i\varepsilon)|\vec{p}\rangle\right] \\ &= \lim_{\varepsilon \to 0}\pm i\varepsilon[g_0(E \pm i\varepsilon) + g_0(E + i\varepsilon)Vg(E + i\varepsilon)]|\vec{p}\rangle \end{aligned} \tag{1.19}$$

For the first term on the right-hand side of the equation, we use the relation

$$\pm i\varepsilon\,g_0(E \pm i\varepsilon)\,|\vec{p}\rangle = |\vec{p}\rangle \tag{1.20}$$

and, using Eq. (1.15), for the second term, we write

$$|\vec{p}\rangle^{(\pm)} = |\vec{p}\rangle + g_0(E \pm i0)\,V\,|\vec{p}\rangle^{\pm} \tag{1.21}$$

This is called the Lippmann–Schwinger (L–S) equation for the scattering states $|p\rangle^{(\pm)}$. The ε limit which is to be performed in $g_0(E \pm i\varepsilon)$ is indicated by the notation $E \pm i0$. The free Green's function has a cut along the positive real axis. The $\varepsilon-$ limit tells us on which side of the cut we have to stay in order to fulfill the boundary condition. The plus sign corresponds to the physical boundary condition. In configuration space representation, Eq. (1.21) written explicitly for the scattering state $|p\rangle^+$ as:

$$\langle \vec{r} \mid \vec{p}\rangle^+ = \langle r|p\rangle + \langle \vec{r}|g_0(E + i0)|\vec{r}'\rangle\langle \vec{r}'|V|\vec{r}''\rangle\langle \vec{r}'' \mid \vec{p}\rangle^+$$

$$\text{or} \quad \psi^{(+)}_{\vec{p}}(\vec{r}) = \phi_{\vec{p}}(\vec{r}) + \int d\vec{r}'d\vec{r}'' \langle \vec{r}|g_0(E + i0)|\vec{r}'\rangle\langle \vec{r}'|V|\vec{r}''\rangle\psi^{(+)}_{\vec{p}}(\vec{r}'') \quad (1.22)$$

For a local potential,

$$\langle \vec{r}'|V|\vec{r}''\rangle = V(\vec{r}')\delta(\vec{r}'' - \vec{r}') \quad (1.23)$$

so in that case

$$\int d\vec{r}'' \langle \vec{r}'|V|\vec{r}''\rangle\psi_p(\vec{r}'') = V(\vec{r}')\psi_{\vec{p}}(\vec{r}') \quad (1.24)$$

For local potentials, the expression (1.22) reduces to:

$$\psi^{(+)}_{\vec{p}}(\vec{r}) = \phi_{\vec{p}}(\vec{r}) + \int d\vec{r}' \langle \vec{r}|g_0(E + i0)|\vec{r}'\rangle V(\vec{r}')\psi^{(+)}_{\vec{p}}(\vec{r}') \quad (1.22')$$

To determine the representation of the free Green's function $\langle \vec{r}|g_0(E + i0)|\vec{r}'\rangle$, we begin by writing:

$$[\nabla_r^2 + p^2] g_0(p, \vec{r}, \vec{r}') = \delta(\vec{r} - \vec{r}'),$$

$$\delta(\vec{r} - \vec{r}') = (2\pi)^{-3} \int \exp[i\,\vec{p}'(\vec{r} - \vec{r}')]d\vec{p}'$$

which gives

$$g_0(p, \vec{r}, \vec{r}') = -(2\pi)^{-3} \int \frac{\exp[i(\vec{p}'.(\vec{r} - \vec{r}')]}{p'^2 - p^2}d\vec{p}' \quad (1.25)$$

The integrand in Eq. (1.25) has poles at $p' = \pm p$. We shall perform the integration by going into the complex p' plane in such a way that $g_0(E, \vec{r}, \vec{r}')$ will lead to an outgoing spherical wave for $r \to \infty$.

To perform the angular integration, we choose spherical polar coordinates such that the vector

$\vec{r} - \vec{r}' \equiv \vec{R}$ coincides with the z-axis. After performing angular integration, we get

$$g_0(p, R) = -(4\pi^2 R)^{-1} \int_{-\infty}^{+\infty} \frac{p' \sin p'R}{p'^2 - p^2} dp'$$

$$\text{or } g_0(p.R) = -(16\pi^2 i R)^{-1} \left\{ \int_{-\infty}^{+\infty} \exp(ip'R) \left[\frac{1}{p' - p} + \frac{1}{p' + p} \right] dp' \right.$$

$$\left. - \int_{-\infty}^{+\infty} \exp(-ip'R) [\frac{1}{p' - p} + \frac{1}{p' + p}] dp' \right\} \tag{1.26}$$

Here, we only examine the first integral on the right-hand side of Eq. (1.26) and analyze this expression by means of contour integration in the complex p'-plane. Calling the integral

$$I_1 = \oint_C \exp(ip'R) \left[\frac{1}{p' - p} + \frac{1}{p' + p} \right] dp'$$

and choosing the contour C in the upper half plane, Im$p' > 0$ by a large semi-circle so that the contribution to the integral from this semi-circle tends to zero as its radius tends to infinity. Out of the four possible choices of performing the integral along the real axis, the one that gives us outgoing spherical wave is as shown in Fig. (1.1).

Thus, the expression:

$$g_0^{(+)}(p, R) = -(4\pi^2 R)^{-1} \oint_P \frac{p' \sin p'R}{p'^2 - p^2} dp' = -\frac{\exp(ipR)}{4\pi R}$$

$$\text{or } g_0^{(+)}(p; \vec{r}, \vec{r}') = -\frac{1}{4\pi} \frac{\exp\{ip|\vec{r} - \vec{r}'|}{|\vec{r} - \vec{r}'|} \tag{1.27}$$

This final expression obtained in Eq. (1.27) is now inserted (with the proviso that $p = \sqrt{2\mu E}$) in Eq. (1.22′) to get

Fig. 1.1 Contour integration on the complex p-plane

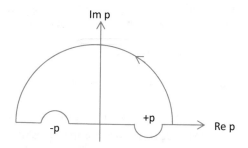

$$\psi_{\vec{p}}^{(+)}(\vec{r}) = \phi_{\vec{p}}(\vec{r}) - \int d\vec{r}' \frac{\mu}{2\pi} \frac{\exp[i\sqrt{2\mu E}\,|\vec{r} - \vec{r}'|]}{|\vec{r} - \vec{r}'|} V(\vec{r}')\,\psi_{\vec{p}}^{(+)}(\vec{r}') \quad (1.28)$$

Let us now ensure whether the kernel of the integral Eq. (1.28) is compact so that the Fredholm theory and all other methods of solving the integral equation can be applied. The simplest way to find out about the compactness is to study the Schmidt norm of the operator $K(\vec{r}, \vec{r}')$, which is defined as

$$\|K\|_S = \left[Tr(K^+ K)\right]^{1/2} = \left[\int\int d\vec{r}\, d\vec{r}'\,|K(\vec{r}, \vec{r}')|^2\right]^{1/2} \quad (1.29)$$

In the integral Eq. (1.28), the kernel is

$$K(\vec{r}, \vec{r}') = -\frac{\mu}{2\pi} \frac{\exp[i\sqrt{2\mu z}\,|\vec{r} - \vec{r}'|]}{|\vec{r} - \vec{r}'|} V(\vec{r}') \quad (1.30)$$

where $z = E + i\varepsilon$. The square of the Schmidt norm is given by

$$
\begin{aligned}
\|K\|_S^2 &= \frac{\mu^2}{4\pi^2} \int\int d\vec{r}\, d\vec{r}' \frac{\exp[-2\text{Im}(\sqrt{2\mu z}\,|\vec{r} - \vec{r}'|]}{|\vec{r} - \vec{r}'|^2} |V(\vec{r}')|^2 \\
&= \frac{\mu^2}{4\pi^2} \int d\vec{R} \frac{\exp[-2\text{Im}\sqrt{2\mu z}\,R]}{R^2} \int d\vec{r}'\,|V(\vec{r}')|^2 \\
&= \frac{\mu^2}{2\pi\,\text{Im}(\sqrt{2\mu z})} \int d\vec{r}'\,|V(\vec{r}')|^2 \quad (1.31)
\end{aligned}
$$

From the expression (1.31), it is clear that Schmidt norm can be finite when (i) the integral over the absolute square of the potential exists, thereby excluding the possibility of Coulomb and hard-core potentials; (ii) the imaginary part of $\sqrt{2\mu z}$ is not zero. The latter condition rules out the possibility in the process of scattering energies, in which we are interested. However, the condition of a finite Schmidt norm is only a sufficient condition and not a necessary one for compactness. Lovelace [2] has shown that despite the divergence of the Schmidt norm, the kernel of the integral equation is compact in the Banach space of continuous bounded functions with continuous bounded derivatives. Thus, for compact kernels, the Lippmann–Schwinger equation has a unique solution.

1.5 The S-Matrix, T-Matrix and Differential Cross Section

The S-matrix provides a link between the scattering states and the experimentally measured data. It is defined as the probability of finding a free wave packet $\phi_b(t)$ in the limit $t \to \infty$ in the scattering state $\psi_a^{(+)}(t)$ (which has developed from the wave packet $\phi_a(t)$ in the infinite past, $t \to -\infty$). Thus,

$$S_{ba} = \lim_{t \to \infty} \langle \phi_b(t) \mid \psi_a^{(+)}(t) \rangle, \tag{1.32}$$

$$= \lim_{t \to \infty} \langle \exp(-iH_0t)\phi_b \mid \exp(-iHt)\psi_a^{(+)} \rangle \tag{1.33}$$

$$= \lim_{t \to \infty} \langle \exp(iHt)\exp(-iH_0t)\phi_b \mid \psi_a^{(+)} \rangle \tag{1.34}$$

Using Eqs. (1.5) and (1.6), we get

$$S_{ba} = \langle \Omega^{(-)}\phi_b \mid \Omega^{(+)}\phi_a \rangle, \tag{1.35}$$

or

$$S_{ba} = \langle \psi_b^{(-)} \mid \psi_a^{(+)} \rangle, \tag{1.36}$$

or

$$S_{ba} = \langle \phi_b | \Omega^{(-)\dagger} \Omega^{(+)} | \phi_a \rangle \tag{1.37}$$

Using momentum eigenstates instead of wave packets, we have

$$S_{\vec{p}'\vec{p}} = \langle \vec{p}' | \Omega^{(-)\dagger} \Omega^{(+)} | \vec{p} \rangle \tag{1.38}$$

In the momentum representation, the *S*-matrix can be considered as an operator

$$S = \Omega^{(-)\dagger} \Omega^{(+)} \tag{1.39}$$

After carrying out the limits (Eq. 1.6), we express the *S*-matrix for momentum states as

$$S_{\vec{p}'\vec{p}} = \lim_{t \to \infty} \exp[i(E'-E)t]\langle p' \mid \vec{p} \rangle^{(+)} \tag{1.40}$$

In general, if the two energies are different, then

$$E' = \frac{p'^2}{2\mu} \neq E = \frac{p^2}{2\mu} \tag{1.41}$$

Expressing the scattering state $|\vec{p}\rangle^{(+)}$ by the resolvent (Eq. 1.19), we write

$$S_{\vec{p}'\vec{p}} = \lim_{t \to \infty} \lim_{\varepsilon \to 0} i\varepsilon \exp[i(E'-E)t]\langle \vec{p}'|G(E+i\varepsilon)|\vec{p} \rangle \tag{1.42}$$

The resolvent satisfies an integral equation similar to the L–S equation [compare Eqs. (1.18a) and (1.19)]. Let us insert Eq. (1.18b) into Eq. (1.18a) to get

$$
\begin{aligned}
G(z) &= G_0(z) + G_0(z)VG_0(z) + G_0(z)VG(z)VG_0(z) \\
&= G_0(z) + G_0(z)[V + VG(z)V]G_0(z) \\
&= G_0(z) + G_0(z)\,T(z)\,G_0(z)
\end{aligned}
\tag{1.43}
$$

We have thus been able to relate the resolvent to the operator $T(z)$, which happens to be less singular than $G(z)$:

$$
T(z) = V + V\,G(Z)\,V
\tag{1.44}
$$

This can be easily seen by writing explicitly the resolvent $G(z)$ in terms of $T(z)$, viz.

$$
\langle \vec{p}'|G(z)|\vec{p}\rangle = \frac{\delta(\vec{p}' - \vec{p})}{z - p^2/2\mu} + \frac{\langle \vec{p}'|T(z)|\vec{p}\rangle}{(z - p'^2/2\mu)\,(z - p^2/2\mu)}
\tag{1.45}
$$

where the kinematic singularities appear explicitly.

The expression (1.45) is now inserted in Eq. (1.43) for the S-matrix and remembering that $z = E + i\varepsilon$, $E = p^2/2\mu$, we get

$$
\begin{aligned}
S_{\vec{p}'\vec{p}} &= \lim_{t\to\infty}\lim_{\varepsilon\to 0}(i\varepsilon)\,\exp([i(E' - E)t]\left\{\frac{\delta(\vec{p}' - \vec{p})}{i\varepsilon} + \frac{\langle \vec{p}'|T(E + i\varepsilon)|\vec{p}\rangle}{(i\varepsilon)\,(E + i\varepsilon - E')}\right\} \\
&= \delta(\vec{p}' - \vec{p}) - \lim_{t\to\infty}\lim_{\varepsilon\to 0}\frac{\exp[i(E' - E)t]}{E' - E - i\varepsilon}\langle \vec{p}'|T(E + i\varepsilon)|\vec{p}\rangle
\end{aligned}
\tag{1.46}
$$

The limits in Eq. (1.46) can be easily carried using the relation

$$
\lim_{t\to\infty}\lim_{\varepsilon\to 0}\frac{\exp(i\omega t)}{\omega - i\varepsilon} = 2\pi i\delta(\omega)
\tag{1.47}
$$

Thus, we finally get

$$
S_{\vec{p}'\vec{p}} = \delta(\vec{p}' - \vec{p}) - 2\pi i\delta(p'^2/2\mu - p^2/2\mu)\langle \vec{p}'|T(E + i0)|\vec{p}\rangle
\tag{1.48}
$$

Thus, in the scattering matrix, the singularities of the resolvent lead to two δ functions: The first one corresponds to non-scattering, i.e., the incident particle going without deflecting, and the second one represents energy conservation. In fact, all the information about the scattering process is contained in the T-matrix, which is directly related to the experimentally measured scattering cross section as:

$$
\frac{d\sigma}{d\Omega} = (2\pi)^4\mu^2\left|\langle \vec{p}'|T(E + i0)|\vec{p}\rangle\right|^2
\tag{1.49}
$$

The state $|\vec{p}\rangle$ defines the direction and energy of the incident particle, and $|\vec{p}'\rangle$ gives the direction of the particle scattered as recorded by the counter. Because of energy conservation, we have

$$\frac{p'^2}{2\mu} = \frac{p^2}{2\mu} = E \tag{1.50}$$

Note that the T-matrix appearing in Eq. (1.49) also defines an off-shell T-matrix for which

$$\frac{p'^2}{2\mu} \neq \frac{p^2}{2\mu} \neq E \tag{1.51}$$

The *T*-matrix
Since the cross section follows directly from the T-matrix, it becomes desirable to have an integral equation for the T-matrix itself. Using Eq. (1.43) and inserting for the resolvent from Eq. (1.18), we have

$$\begin{aligned}
G(z) &= G_0(z) + G_0(z)T(z)G_0(z) \\
&= G_0(z) + G_0(z)VG_0(z) + G_0(z)VG_0(z)T(z)G_0(z)
\end{aligned} \tag{1.52}$$

From this equation, it follows that

$$T(z) = V + VG_0(z)\,T(z)$$

or

$$T(z) = V + T(z)G_0(z)V \tag{1.53a, b}$$

These equations, written explicitly in representation space, are the integral equations for T-matrix. It may be noted that all these integral equations, like that of L–S equation, have the same kernel.

1.6   The Unitarity of *S*-Matrix and Optical Theorem

There exists a conservation law for the probability flux if the given potential is Hermitian. As a consequence, the S-matrix is required to be unitary, i.e.,

$$S^+S = \mathbb{1} \tag{1.54}$$

One can derive the consequence of this relation for the T-matrix starting from Eq. (1.48).

However, one can also start using Eq. (1.53a) by multiplying by $T^{-1}(z)$ from the right and by V^{-1} from the left to get the equation:

$$T^{-1}(z) = V^{-1} - G_0(z) \tag{1.55}$$

As $G_0(z)$ has a cut along the positive real energy axis so has T(z) and $T^{-1}(z)$. Writing Eq. (1.55) for $z = E + i\varepsilon$,

$$T^{-1}(E + i\varepsilon) = V^{-1} - G_0(E + i\varepsilon) \tag{1.56a}$$

and for $z = E - i\varepsilon$,

$$T^{-1}(E - i\varepsilon) = V^{-1} - G_0(E - i\varepsilon) \tag{1.56b}$$

Subtracting these two equations, we have

$$T^{-1}(E + i\varepsilon) - T^{-1}(E - i\varepsilon) = -G_0(E + i\varepsilon) + G_0(E - i\varepsilon)$$

And multiplying this equation by $T(E + i\varepsilon)\,T(E - i\varepsilon)$ on both sides, we get

$$T(E - i\varepsilon) - T(E + i\varepsilon) = T(E + i\varepsilon)[-G_0(E + i\varepsilon) + G_0(E - i\varepsilon)]\,T(E - i\varepsilon) \tag{1.57}$$

Using the first resolvent equation, the difference in parentheses reduces to:

$$-G_0(E + i\varepsilon) + G_0(E - i\varepsilon) = 2i\varepsilon G_0(E + i\varepsilon)G_0(E - i\varepsilon)$$
$$= 2i\varepsilon[(E - H_0)^2 + \varepsilon^2] \tag{1.58}$$

In the limit $\varepsilon \to 0$, we get the unitarity relation:

$$T(E - i0) - T(E + i0) = 2\pi i T(E + i0)\delta(E - H_0)T(E - i0),$$

or

$$T(E + i0) - T^+(E + i0) = -2\pi i T(E + i0)\delta(E - H_0)T^+(E + i0) \tag{1.59}$$

Expressed in momentum representation, this expression for scattering in the forward direction reduces to:

$$\text{Im}\langle \vec{p}|T(E + i0)|\vec{p}\rangle = -\pi \int d\vec{p}'\delta(E - p'^2/(2\mu))\left|\langle \vec{p}|T(E + i0)|\vec{p}'\rangle\right|^2$$
$$= -\pi \mu p \int d\hat{p}'\left|\langle \vec{p}|T(E + i0)|\vec{p}'\rangle\right|^2, \quad |\vec{p}| = \left|\vec{p}' = \sqrt{2\mu E}\right|$$
$$= -\frac{p}{16\pi^3\mu}\sigma_{\text{tot}} \tag{1.60}$$

This is the well-known optical theorem which relates the imaginary part of the forward amplitude to the total cross section.

Chapter 2
Scattering Theory of Three-Particle System

2.1 Introduction

In this chapter, we present a brief and simple formulation of Faddeev's three-particle scattering theory. Recognizing the basic complexities involved while extending the two-body Lippmann–Schwinger equation to the three-particle case with pair-wise potentials, Faddeev [7] showed how the problems of disconnectedness of the scattering matrix in the three-body kernel can be resolved by expressing the three-particle T-matrix as a sum of three terms where each term is expressed in terms of the other two and the two body potentials are replaced by the corresponding T-matrices for the two particles. He also successfully tackled the problem of non-uniqueness of the boundary conditions for different channels encountered in the usual Lippmann–Schwinger equation for three-particle scattering. Faddeev [7] undertook a detailed investigation to study the analytic behavior of the scattering amplitude of the resulting integral equations to show that these equations have completely continuous(compact) kernel and has unique solutions at scattering energies. In the last sub-section, we discuss the solution of three-particle system using Schrodinger equation with separable pair-wise potentials and derive the integral equation for the scattering amplitude of a particle by the bound state of the other two. The idea is to demonstrate how the problems of disconnectedness of the scattering matrix and the non-uniqueness of the boundary condition are resolved in this approach.

V. S. Bhasin and I. Mazumdar, *Few Body Dynamics, Efimov Effect and Halo Nuclei*, SpringerBriefs in Physics, https://doi.org/10.1007/978-3-030-56171-0_2

2.2 Three-Particle Hamiltonian, Channel Hamiltonian, Matrix-Elements for the Green's Functions and T-Matrix

Let us now extend the formulation of two-body problem studied above to the case of three-particle scattering. Consider the Hamiltonian for three particles of masses m_1, m_2 and m_3 interacting through pair-wise potentials, i.e.,

$$H = H_0 + V, \tag{2.1}$$

where H_0 is the kinetic energy operator of the system,

$$H_0 = \frac{p_1^2}{2m_1} + \frac{p_2^2}{2m_2} + \frac{p_3^2}{2m_3},$$

$$\text{and} \quad V = \sum_{i=1}^{3} V_i = V_{23} + V_{31} + V_{12},$$

$$\text{with} \quad V_1 = V_{23}, \ V_2 = V_{31} \ \text{and} \ V_3 = V_{12} \tag{2.1a–c}$$

The Cartesian space coordinates, $\vec{r}_1, \vec{r}_2, \vec{r}_3$ of the three particles are written in terms of the Jacobi coordinates:

$$\vec{\eta}_1 = \vec{r}_2 - \vec{r}_3$$

$$\vec{\rho}_1 = \vec{r}_1 - \frac{m_2\vec{r}_2 + m_3\vec{r}_3}{m_2 + m_3}$$

$$\vec{R} = \frac{m_1\vec{r}_1 + m_2\vec{r}_2 + m_3\vec{r}_3}{m_1 + m_2 + m_3} \tag{2.2}$$

where particles 2 and 3 appear explicitly forming a two-particle sub-system. The momentum coordinates $\vec{p}_1, \vec{p}_2$ and $\vec{p}_3$ are correspondingly transformed as

$$\vec{q}_1 = \frac{m_3\vec{p}_2 - m_2\vec{p}_3}{m_2 + m_3}$$

$$\vec{k}_1 = \frac{(m_2 + m_3)\vec{p}_1 - m_1(\vec{p}_2 + \vec{p}_3)}{m_1 + m_2 + m_3}$$

$$\vec{P} = \vec{p}_1 + \vec{p}_2 + \vec{p}_3 \tag{2.3}$$

In Jacobi coordinates, the Hamiltonian operator is expressed as

$$H = \frac{P^2}{2(m_1 + m_2 + m_3)} + \frac{q_1^2}{2\mu_1} + \frac{k_1^2}{2\hat{\mu}_1} + \sum_{i=1}^{3} V_i \tag{2.4}$$

where the reduced masses are given by

$$\mu_1 = \frac{m_2 m_3}{m_2 + m_3} \quad \text{and} \quad \widehat{\mu}_1 = \frac{m_1(m_2 + m_3)}{m_1 + m_2 + m_3} \tag{2.5}$$

The physical interpretation of the kinetic energy part of the Hamiltonian is that the first term is the kinetic energy of the CM, the second term corresponds to the relative kinetic energy of particles 2 and 3 in their center of mass system, while the third term accounts for the relative kinetic energy of particle 1 and the composite system (2, 3). Thus, in the CM system, the total Hamiltonian can be written as:

$$H = \frac{q_i^2}{2\mu_i} + \frac{k_i^2}{2\widehat{\mu}_i} + \sum_{j=1}^{3} V_j \quad (i = 1, 2, \text{or } 3) \tag{2.6}$$

In the three-body problem, two-particle sub-systems have a crucial role for which we introduce a channel Hamiltonian, H_i which is defined as

$$H_i = \frac{q_i^2}{2\mu_i} + \frac{k_i^2}{2\widehat{\mu}_i} + V_i \tag{2.7}$$

and is related to the total Hamiltonian H as

$$H = H_i + \bar{V}_i$$

$$\text{where} \quad \bar{V}_i = V - V_i = \sum_{j \neq i}^{3} V_j \tag{2.8, 2.9}$$

It may be useful to include also the breakup channel (i.e., when all the three particles are free) by extending the value of index $i = 0$ (in addition to 1, 2 or 3). In that case $V_0 = 0$ and Eqs. (2.7) and (2.8) become also valid for $i = 0$. Along with the channel Hamiltonian, we also introduce channel resolvent Green's function G_i defined as:

$$G_i(z) \equiv (z - H_i)^{-1} \tag{2.10}$$

Thus, written explicitly, the matrix element of the channel resolvent would appear as

$$\left\langle \vec{q}_i, \vec{k}_i \middle| G_i(z) \middle| \vec{q}_i', \vec{k}_i' \right\rangle = \delta(\vec{k}_i - \vec{k}_i') \langle \vec{q}_i | G_i(z - k_i^2/2\widehat{\mu}_i) | \vec{q}_i' \rangle \tag{2.11}$$

These matrix elements are similar to the two-body matrix elements of two particle Green's functions introduced earlier [cf. Eq. (1.14)]. The Green's function G_i carries the channel index 'i' because different potentials V_i can act in different channels. The $\delta-$ function accounts for the fact that G_i contains only the interaction of the sub-system 'i' and particle 'i' is free. The argument of G_i is the energy of the sub-system 'i', obtained by subtracting the kinetic energy of the particle 'i' from the free energy z. For the Green's function G_0 for the three particles, we have

$$\left\langle \vec{q}_i, \vec{k}_i \middle| G_0(z) \middle| \vec{q}_i', \vec{k}_i' \right\rangle = \frac{\delta(\vec{k}_i - \vec{k}_i') \, \delta(\vec{q}_i - \vec{q}_i')}{z - k_i^2/2M_i - q_i^2/2\mu_i} \tag{2.12}$$

The expression for the T-matrix T_i for the two-particle sub-system in three-particle Hilbert space can also be written similar to Eq. (2.11) as

$$\left\langle \vec{q}_i, \vec{k}_i \middle| T_i(z) \middle| \vec{q}_i', \vec{k}_i' \right\rangle = \delta(\vec{k}_i - \vec{k}_i') \, \langle \vec{q}_i | T_i(z - k_i^2/2\hat{\mu}_i) | \vec{q}_i' \rangle \tag{2.13}$$

2.3 Boundary Conditions and Moller Operators

The discussion of boundary conditions presented in the case of two-particle scattering can be easily extended to the three-particle case. Defining the time development of a three-particle wave packet by

$$\psi_{ij}^{(+)}(t) = \exp(-iHt) \, \psi_{ij}^{(+)} \tag{2.14}$$

and the reference wave packets which develop according to

$$\phi_{ij}(t) = \exp(-iH_i t) \, \phi_{ij} \tag{2.15}$$

The wave packet ϕ_{ij} describes the free motion of the particle 'i' ($i = 1, 2, 3$) relative to the other two particles which are in their jth bound state. We demand that the norm of the difference vanishes in the distant past leading to the representation of scattering states

$$\psi_{ij}^{(+)} = s - \lim_{t \to \mp\infty} \exp(iHt) \exp(-iH_i t)\phi_{ij} \equiv \Omega_i^{(\pm)} \phi_{ij}$$

where Moller operators are defined as

$$\Omega_i^{(\pm)} = s - \lim_{t \to \mp} \exp(-iHt) \, \exp(-iH_i t), \quad i = 1, 2, 3 \tag{2.16}$$

Here we also include the $t \to +\infty$ which is required for the construction of S-matrix.

The fact that a separate Moller operator is needed for every partition 'i' itself shows that the three-particle problem is much more difficult to handle. The domain of Moller operators defined above is the space of channel states

$$\left|\psi_{ij}^{(\pm)}\right\rangle = \Omega_{ij}^{i\ (\pm)}\left|\phi_{ij}\right\rangle \tag{2.17}$$

Transition from time limit to Euler limit is also possible for three-particle scattering

$$\left|\psi_{ij}^{(\pm)}\right\rangle = \lim_{\varepsilon \to 0} \pm i\varepsilon G(E \mp i\varepsilon)\left|\phi_{ij}\right\rangle \tag{2.18}$$

where G is the full resolvent or Green's function satisfying the relation

$$\begin{aligned} G(z) &= G_i(z) + G_i(z)\,\bar{V}_i\,G(z) \\ &= G_i(z) + G(z)\,\bar{V}_i(z)\,G_i(z) \end{aligned} \tag{2.19}$$

in terms of the channel resolvent G_i.

2.4 Difficulties with the 3-Particle Lippmann–Schwinger Equation

2.4.1 Non-uniqueness of the Boundary Conditions

Operating Eq. (2.19) on the state $\left|\phi_{ij}\right\rangle$ we have

$$\begin{aligned} \lim_{\varepsilon \to 0} \pm i\varepsilon G(E \pm i\varepsilon)\left|\phi_{ij}\right\rangle = \lim_{\varepsilon \to 0} \pm i\varepsilon \{ & G_i^{(\pm)}(E \pm i\varepsilon) \\ & + G_i^{(\pm)}(E \pm i\varepsilon)\,\bar{V}_i\,G(E \pm i\varepsilon)\}\left|\phi_{ij}\right\rangle \end{aligned} \tag{2.20}$$

[Note: the letter i before ε stands for imaginary and need not be confused with the index like ij].

We now use Eq. (2.18) and the relation

$$\lim_{\varepsilon \to 0} \pm i\varepsilon\, G_i^{(\pm)}(E \pm i\varepsilon)\left|\phi_{ij}\right\rangle = \left|\phi_{ij}\right\rangle \tag{2.21}$$

to get the L–S equation

$$\left|\psi_{ij}^{(\pm)}\right\rangle = \left|\phi_{ij}\right\rangle + G_i^{(\pm)}(E \pm i0)\,\bar{V}_i\left|\psi_{ij}^{(\pm)}\right\rangle \tag{2.22}$$

Solution to the L–S equation is not uniquely determined, because we shall find the homogeneous equation

$$\left|\psi\right\rangle = G_i(E \pm i0)\,\bar{V}_i\left|\psi\right\rangle \tag{2.23}$$

has solution for energies in the scattering region.

Consider, for example, another channel state

$$|\phi_{jn}\rangle \quad \text{for } j \neq i$$

For such a state

$$\lim_{\varepsilon \to 0} \pm i\varepsilon \, G_i^{(\pm)}(E \pm i\varepsilon)|\phi_{jn}\rangle = 0 \qquad (2.24)$$

because the state $|\phi_{jn}\rangle$ *for* $j \neq i$ is not an eigen-state of H_i. Therefore, we get

$$\left|\psi_{jn}^{(\pm)}\right\rangle = G_i(E + i0)\,\bar{V}_i\left|\psi_{jn}^{(\pm)}\right\rangle \qquad (2.25)$$

This homogeneous equation which belongs to the inhomogeneous L–S equation has non-trivial solutions of the scattering states in channels, $j \neq i$.

2.4.2 Disconnectedness of the Kernel

Another difficulty one encounters is that the kernel of L–S Eq. (2.22) does not have a finite Schmidt norm. The kernels are also not compact and therefore standard methods of the theory of integral equations can not be applied. Written explicitly in momentum space

$$\langle \vec{p}_1 \vec{p}_2 \vec{p}_3 | G_0(z) \, \bar{V}_i | \vec{p}_1' \vec{p}_2' \vec{p}_3' \rangle = (z - \sum_{i=1}^{3} p_i^2/2m_i)^{-1} \sum_{j \neq i} \delta(\vec{k}_j - \vec{k}_j')\langle \vec{q}_j | \bar{V}_j | \vec{q}_j' \rangle$$

$$(2.26)$$

In Eq. (2.26), the δ-function belonging to the total CM motion is omitted on the right-hand side. However, δ-functions associated with the two-body potentials signify that the third particle remains free when the other two particles are interacting. Thus, the kernel $G_0\bar{V}$ gets disconnected. As a result, the kernel of L–S equation is no longer square integrable. The disconnected parts of the kernel can be diagrammatically shown as in Fig. 2.1:

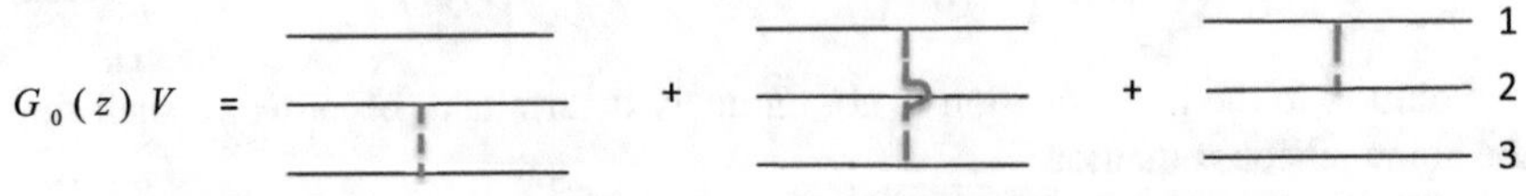

Fig. 2.1 Kernel of the three particle Lippmann-Schwinger equation shown diagrammatically. The red dashed lines represent the interaction between the pairs while the third particle remains free

2.5   The Faddeev Equations

In order to remove these difficulties, Faddeev's starting point is to extend the equation for two-body T-matrix to three-particle scattering:

$$T(z) = V + V\,G(z)\,V \qquad (2.27)$$

This operator equation is the formal analog of the two-particle T-operator (Eq. 1.44). However, Eq. (2.27) is not as directly related to the scattering cross section as in the two-particle case. In analogy to Eq. (1.43), we use the equation

$$G(z) = G_0(z) + G_0(z)T(z)G_0(z) \qquad (2.28)$$

to write the equation for $T(z)$ in the operator form as

$$
\begin{aligned}
T(z) &= V + V G_0(z)T(z) \\
     &= V + T(z)G_0(z)V
\end{aligned}
\qquad (2.29a, b)
$$

similar to the two-particle case [cf. Eqs. (1.53a) and (1.53b). These integral equations for the operator suffer from the same difficulties as the Lippmann–Schwinger equation since they have the same kernel.

Faddeev proposed a novel but simple idea to split the T-matrix in Eq. (2.29a) into three parts:

$$T(z) = T^{(1)}(z) + T^{(2)}(z) + T^{(3)}(z) \qquad (2.30)$$

where using Eq. (2.27a), we write

$$T^{(i)} = V_i + V_i G_0 T \quad (i = 1, 2, 3) \qquad (2.31)$$

Here we are omitting the z-dependence of the operators. Writing Eq. (2.31) explicitly for $i = 1$, we have

$$T^{(1)} = V_1 + V_1 G_0 (T^{(1)} + T^{(2)} + T^{(3)}) \qquad (2.32)$$

Now let us bring the term appearing on the right-hand side of Eq. (2.32) to the left to get

$$
\begin{aligned}
&T^{(1)} - V_1 G_0 T^{(1)} = (1 - V_1 G_0)T^{(1)} = V_1 + V_1 G_0\big(T^{(2)} + T^{(3)}\big) \\
&\text{or}\quad (1 - V_1 G_0)T^{(1)} = V_1 + V_1 G_0\big(T^{(2)} + T^{(3)}\big) \\
&\text{or}\quad T^{(1)} = (1 - V_1 G_0)^{-1}V_1 + (1 - V_1 G_0)^{-1}V_1 G_0\big(T^{(2)} + T^{(3)}\big) \\
&\text{i.e.,}\quad T^{(1)} = T_1 + T_1 G_0\big(T^{(2)} + T^{(3)}\big)
\end{aligned}
\qquad (2.33)
$$

where for the two-body T-matrix, we write $T_1 = (1 - V_1 G_0)^{-1} V_1$ [cf., Eq. (1.53a)]. Equations similar to Eq. (2.31) for $T^{(2)}$ and $T^{(3)}$ can also be obtained by using Eq. (2.29). We thus write the three components in the matrix form as

$$
\begin{pmatrix} T^{(1)} \\ T^{(2)} \\ T^{(3)} \end{pmatrix} = \begin{pmatrix} T_1 \\ T_2 \\ T_3 \end{pmatrix} + \begin{pmatrix} 0 & T_1 & T_1 \\ T_2 & 0 & T_2 \\ T_3 & T_3 & 0 \end{pmatrix} G_0 \begin{pmatrix} T^{(1)} \\ T^{(2)} \\ T^{(3)} \end{pmatrix} \tag{2.34}
$$

In Eq. (2.34), which we call the Faddeev equation, the new features that have appeared are: (i) instead of two body potentials, we now have two-particle T-matrices which are to be understood as operators in three-particle space. The T_i's enter into the Faddeev equation as off-shell quantities [see, e.g., Eq. (2.13)]. Thus, we have got a mathematical formulation where the two-particle scattering amplitudes, which are more closely related to experiment, enter into the three-particle amplitude containing more information; (ii) the kernel of the three-body T-matrix gets connected-the dangerous δ-functions no longer appear in the kernel. This can be checked by iteration of the kernel in Eq. (2.34) as

$$
\begin{pmatrix} T^{(1)} \\ T^{(2)} \\ T^{(3)} \end{pmatrix} = \begin{pmatrix} T_1 \\ T_2 \\ T_3 \end{pmatrix} + \begin{pmatrix} T_1 G_0 (T_2 + T_3) \\ T_2 G_0 (T_3 + T_1) \\ T_3 G_0 (T_1 + T_2) \end{pmatrix}
$$
$$
+ \begin{pmatrix} T_1 G_0 (T_2 + T_3) & T_1 G_0 T_3 & T_1 G_0 T_2 \\ T_2 G_0 T_3 & T_2 G_0 (T_3 + T_1) & T_2 G_0 T_1 \\ T_3 G_0 T_2 & T_3 G_0 T_1 & T_3 G_0 (T_1 + T_2) \end{pmatrix} G_0 \begin{pmatrix} T^{(1)} \\ T^{(2)} \\ T^{(3)} \end{pmatrix} \tag{2.35}
$$

The iterated terms contain only operator products $T_i G_0 T_j$ with $i \neq j$ and therefore the kernel contains only terms where all three particles are connected together. For example, as shown in Fig. 2.2, the new kernel K_{12}^2 is just

The work by Faddeev and Lovelace has shown that the operator K_{ij}^2 is compact and with certain requirements on the interaction potentials and can even be square integrable for all but physical values of z. The limit as z becomes real has been investigated in detail by Faddeev who shows that for real z, the fifth power of kernel is a compact operator in a Banach space. This ensures that the solution of Faddeev equations is unique.

Having discussed Faddeev equations in terms of T-matrices, similar equations can also be written for the total Green's operators or for the wave functions. Thus,

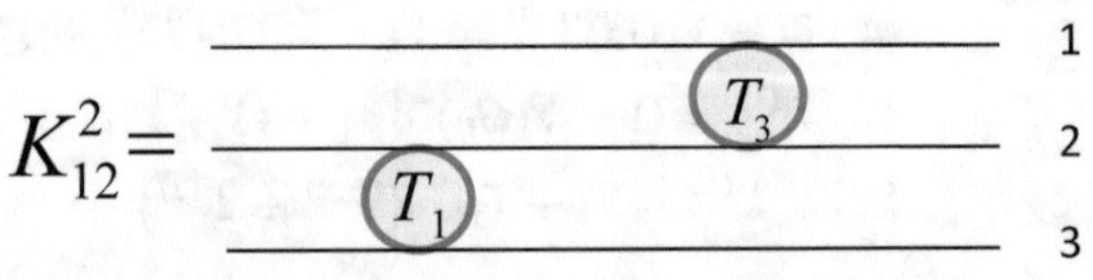

Fig. 2.2 Graphical representation of a typical term of the iterated kernel

we have

$$G = G_0 + GVG_0$$

or

$$G = G_0 + G_0TG_0$$

$$\text{or} \quad G = G_0 + \sum_{i=1}^{3} G_0 T^{(i)} G_0 \qquad (2.36, 2.37)$$

Defining the components

$$G^{(i)}(z) = G_0(z) T^{(i)} G_0(z) \qquad (2.38)$$

The Green's operators $G^{(i)}(z)$ then satisfy the Faddeev equations

$$\begin{pmatrix} G^{(1)} \\ G^{(2)} \\ G^{(3)} \end{pmatrix} = \begin{pmatrix} G_1 - G_0 \\ G_2 - G_0 \\ G_3 - G_0 \end{pmatrix} + G_0 \begin{pmatrix} 0 & T_1 & T_1 \\ T_2 & 0 & T_2 \\ T_3 & T_3 & 0 \end{pmatrix} \begin{pmatrix} G^{(1)} \\ G^{(2)} \\ G^{(3)} \end{pmatrix}$$

or

$$G^{(k)}(z) = G_k(z) - G_0(z) + \sum_{j=1}^{3} G_0(z) M_{kj} G^{(k)}(z), \qquad (2.39a, b)$$

where M_{kj} is the Faddeev (3×3) matrix appearing in Eq. (2.39a).

2.6 Faddeev Equations for Scattering States

Let us now see how the problem of uniqueness is achieved in Faddeev's theory. Starting from Eq. (2.18) for the scattering state $\left| \psi_{ij}^{(\pm)} \right\rangle$ we have

$$\left| \psi_{ij}^{(\pm)} \right\rangle = \lim_{\varepsilon \to 0} \pm i\varepsilon G(E \pm i\varepsilon) \left| \phi_{ij} \right\rangle \qquad (2.40)$$

Using (2.37) and (2.38), we get

$$\left| \psi_{ij}^{(\pm)} \right\rangle = \lim_{\varepsilon \to 0} \pm i\varepsilon G_0(E \pm i\varepsilon) \left| \phi_{ij} \right\rangle + \lim_{\varepsilon \to 0} \pm i\varepsilon \sum_{k=1}^{3} G^{(k)} \left| \phi_{ij} \right\rangle \qquad (2.41)$$

Introducing the definitions

$$\left|\chi_{kij}^{(\pm)}\right\rangle = \lim_{\varepsilon \to 0} \pm i\varepsilon G_k(E \pm i\varepsilon)\left|\phi_{ij}\right\rangle \quad k = 0, 1, 2, 3$$

and

$$\left|\psi_{ij}^{(\pm)}\right\rangle^{(k)} = \lim_{\varepsilon \to 0} \pm i\varepsilon G^{(k)}(E \pm i\varepsilon)\left|\phi_{ij}\right\rangle \quad k = 1, 2, 3 \qquad (2.42a, b)$$

We obtain the scattering state split into the components

$$\left|\psi_{ij}^{(\pm)}\right\rangle = \left|\chi_{0ij}^{(\pm)}\right\rangle + \sum_{k=1}^{3} \left|\psi_{ij}^{(\pm)}\right\rangle^{(k)} \qquad (2.43)$$

Using the Green's operator Eq. (2.39b) on Eq. (2.42b), we have

$$\left|\psi_{ij}^{(\pm)}\right\rangle^{(k)} = \lim_{\varepsilon \to 0} \pm i\varepsilon \Bigg\{ G_k(E \pm i\varepsilon) - G_0(E \pm i\varepsilon)$$
$$+ \sum_{l=1}^{3} G_0(E \pm i\varepsilon) M_{kl}(E \pm i\varepsilon) G^{(l)}(E \pm i\varepsilon) \Bigg\} \left|\phi_{ij}\right\rangle \qquad (2.44)$$

We now use Eq. (2.42) once again to get

$$\left|\psi_{ij}^{(\pm)}\right\rangle^{(k)} = \left|\chi_{ij}^{(\pm)}\right\rangle^{(k)} - \left|\chi_{ij}^{(\pm)}\right\rangle^{(0)} + \sum_{l=1}^{3} G_0(E \pm i\varepsilon) M_{kl}(E \pm i\varepsilon) \left|\psi_{ij}^{(\pm)}\right\rangle^{(l)} \qquad (2.45)$$

We have a bound pair in the entrance channel, i.e.; $i \neq 0$, we shall have

$$\left|\chi_{ij}^{(\pm)}\right\rangle^{(k)} = \delta_{ki}\left|\phi_{ij}\right\rangle, \quad k = 0, 1, 2, 3, \qquad (2.46)$$

For $i = 0$, the scattering state evolves from a free state $|\phi_0\rangle$ we thus get

$$\left|\chi_{k0}^{(\pm)}\right\rangle = \begin{cases} \lim_{\varepsilon \to 0} \pm i\varepsilon G_0(E \pm i\varepsilon)|\phi_0\rangle = |\phi_0\rangle = 0, & k = 0 \\ \lim_{\varepsilon \to 0} \pm i\varepsilon G_k(E + i\varepsilon)|\phi_0\rangle = \Omega_k^{(\pm)}|\phi_0\rangle, & k \neq 0 \end{cases} \qquad (2.47)$$

where $\Omega_i^{(\pm)}$ is the two-particle Moller operator in three-particle space. With Eqs. (2.45) and (2.46), the Faddeev scattering state equations when there is a bound pair in the entrance channel ($k \neq 0$) are written as:

$$\left|\psi_{ij}^{(\pm)}\right\rangle = \sum_{k=1}^{3} \left|\psi_{ij}^{(\pm)}\right\rangle^{(k)},$$

where

$$\left|\psi_{ij}^{(\pm)}\right\rangle^{(k)} = \delta_{ki}|\phi_{ij}\rangle + \sum_{l=1}^{3} G_0(E+i\varepsilon)M_{kl}(E+i0)\left|\psi_{ij}^{(\pm)}\right\rangle^{(l)} \tag{2.48}$$

For the case when there are three free particles in the entrance channel, i.e., $k = 0$ are given by

$$\left|\psi_0^{(\pm)}\right\rangle = |\phi_0\rangle + \sum_{i=1}^{3} \left|\psi_0^{(\pm)}\right\rangle^{(i)},$$

where

$$\left|\psi_0^{(\pm)}\right\rangle^{(i)} = \left|\varphi^{(\pm)}\right\rangle^{(i)} + \sum_{j=1}^{3} G_0(E \pm i\varepsilon)M_{ij}\left|\psi_0^{(\pm)}\right\rangle^{(j)},$$

with

$$\left|\varphi^{(\pm)}\right\rangle^{(i)} = \Omega_i^{(\pm)}|\phi_0\rangle - |\phi_0\rangle, \tag{2.49, 2.50}$$

which represents a spherical outgoing or incoming wave respectively.

The formulation presented so far is quite general. Let us consider one particular case which describes a situation in which initially particle 1 is free and is scattered by the pair (2, 3) which is bound.

Also simplifying the notation, we write the total Hamiltonian

$$H = H_1 + V_1, \text{ with } H_1 = H_0 + V^{(1)} = H_0 + V_{23}, V_1 = V - V^{(1)}$$

where $H_1\phi_{1n} = E_{1n}\phi_{1n}$ and the eigen state $|\psi_{1n}\rangle = \left|\psi^{(1)}\right\rangle + \left|\psi^{(2)}\right\rangle + \left|\psi^{(3)}\right\rangle$

With

$$\begin{pmatrix} \left|\psi^{(1)}\right\rangle \\ \left|\psi^{(2)}\right\rangle \\ \left|\psi^{(3)}\right\rangle \end{pmatrix} = \begin{pmatrix} |\phi_{1n}\rangle \\ 0 \\ 0 \end{pmatrix} + G_o(z)\begin{pmatrix} 0 & T_1(z) & T_1(z) \\ T_2(z) & 0 & T_2(z) \\ T_3(z) & T_3(z) & 0 \end{pmatrix}\begin{pmatrix} \left|\psi^{(1)}\right\rangle \\ \left|\psi^{(2)}\right\rangle \\ \left|\psi^{(3)}\right\rangle \end{pmatrix}, \tag{2.51}$$

the index n contains the additional information on bound states, spin, i-spin, etc.

Similarly, for the break-up channel, we have

$$|\psi_{n0}\rangle = |\phi_{n0}\rangle + \left|\psi_{n0}^{(1)}\right\rangle + \left|\psi_{n0}^{(2)}\right\rangle + \left|\psi_{n0}^{(3)}\right\rangle \quad with \ (z = E_{n0} + i0)$$

$$\begin{pmatrix} \left|\psi_{n0}^{(1)}\right\rangle \\ \left|\psi_{n0}^{(2)}\right\rangle \\ \left|\psi_{n0}^{(3)}\right\rangle \end{pmatrix} = \begin{pmatrix} \left|\phi_{n0}^{(1)}\right\rangle - |\phi_{n0}\rangle \\ \left|\phi_{n0}^{(2)}\right\rangle - |\phi_{n0}\rangle \\ \left|\phi_{n0}^{(3)}\right\rangle - |\phi_{n0}\rangle \end{pmatrix} - G_0\begin{pmatrix} 0 & T_1(z) & T_1(z) \\ T_2(z) & 0 & T_2(z) \\ T_3(z) & T_3(z) & 0 \end{pmatrix}\begin{pmatrix} \left|\psi_{n0}^{(1)}\right\rangle \\ \left|\psi_{n0}^{(2)}\right\rangle \\ \left|\psi_{n0}^{(3)}\right\rangle \end{pmatrix} \tag{2.52}$$

In order to write these equations explicitly, let us adopt the momentum space representation and consider a simplified case where all the three particles are identical so that all the states $|\psi^{(i)}\rangle$s coincide and we shall have an integral equation for a single state $|\psi_n\rangle$:

$$\left\langle \vec{q}, \vec{k} \mid \psi_n \right\rangle \equiv \psi_n(\vec{q}, \vec{k}) = \left\langle \vec{q}, \vec{k} \mid \phi_n \right\rangle - \frac{m}{q^2 + 3k^2/4 - mz} \int \left[\langle \vec{q} | T \left(z - \frac{3k^2}{4m} \right) \right.$$

$$\left| -\frac{1}{2}\vec{k} - \vec{k}' \right\rangle \left\langle \vec{k} + \frac{1}{2}\vec{k}', \vec{k}' \mid \psi_n \right\rangle + \langle \vec{q} | T \left(z - \frac{3k^2}{4m} \right)$$

$$\left. \left| \frac{1}{2}\vec{k} + \vec{k}' \right\rangle \left\langle -\vec{k} - \frac{1}{2}\vec{k}', \vec{k}' \mid \psi_n \right\rangle \right] d\vec{k}' \quad \text{with } z = E_n + i0 \qquad (2.53)$$

Note that for identical particles
$m_1 = m_2 = m_3 = m;$ and in the CM system $\vec{p}_1 + \vec{p}_2 + \vec{p}_3 = 0$ so that for $\vec{q}_1 \equiv \vec{q} = \frac{\vec{p}_2 = \vec{p}_3}{2}$ and $\vec{k}_1 \equiv \vec{k} = \frac{2\vec{p}_1 + \vec{p}_1}{3} = \vec{p}_1$. The total kinetic energy for the three particles is, $KE = \frac{q^2}{2\mu_1} + \frac{k^2}{2\bar{\mu}_1} = \frac{q^2}{m} + \frac{3k^2}{4m}$

To simplify the first integral in Eq. (2.53), we use the δ- functions with the two-body T-matrices [cf. Eq. (2.13)] and the overall δ- function for the 3-body CM system. Thus, the first term in the integral on the right-hand side which is

$$\int \langle \vec{q}_{23} | T \left(z - \frac{3k^2}{4m} \right| \vec{q}'_{23} \rangle \langle \vec{q}'_{12}, \vec{k}'_3 \mid \psi_n \rangle d\vec{k}'_3$$

can be simplified by using $\vec{q}'_{12} = \frac{\vec{p}'_1 - \vec{p}'_2}{2} = \vec{p}_1 + \vec{p}'_3/2 \equiv \vec{k} + \vec{k}'/2$ and $\vec{p}'_3 \equiv \vec{k}'$. In exactly the same way, the second term of the integral is written. The inhomogeneous term

$$\left\langle \vec{q}, \vec{k} \mid \phi_n \right\rangle \equiv \phi_n(\vec{q}, \vec{k}) = \varphi(\vec{q}) \, \delta(\vec{k} - \vec{k}_0), \qquad (2.54)$$

where $\varphi(\vec{q})$ is the function describing the bound state of the two particles and $\delta(\vec{k} - \vec{k}_0)$ represents the plane wave of the incident particle; $E_n = 3k_0^2/4m - i0$. The function Ψ_{sym} is expressed as

$$\Psi_{\text{sym}} = \psi_n(\vec{q}_{23}, \vec{k}_1) + \psi_n(\vec{q}_{31}, \vec{k}_2) + \psi_n(\vec{q}_{12}, \vec{k}_3) = 3 \, \text{sym}_{\vec{q}, \vec{k}} [\psi_n(\vec{q}_{23}, \vec{k}_1)] \quad (2.55)$$

The symbol 'sym' denotes the symmetry with respect to the variables written in the subscript.

In the case of breakup reaction corresponding to the motion of all three particles, we have

$$\phi_n(\vec{q}, \vec{k}) = m \frac{\langle \vec{q} | T(q_0^2/m + i0) | \vec{q}_0 \rangle}{q^2 - q_0^2 - i0} \delta(\vec{k} - \vec{k}_0), \quad \text{where } E_n = \frac{q_0^2}{m} + \frac{3k_0^2}{4m},$$

$$\Psi_{\text{sym}} = \text{sym}_{q,k;q_0,k_0} [\delta(\vec{q}_{23} - \vec{q}_{23}^{\,0}) \delta(\vec{k}_1 - \vec{k}_1^0) + 3\psi_n(\vec{q}_{23}, \vec{k}_1; \vec{q}_{23}^{\,0}, \vec{k}_1^0)] \quad (2.56)$$

where for ψ_n, Eq. (2.52) can be explicitly written in momentum space in the same way as shown in Eq. (2.53).

2.7 Three-Body Problem with Separable Potentials

As a comparative study with Faddeev's theory of three-particle scattering, we consider the scattering of a particle by the bound state of other two through the exact solution of three-particle Schrodinger equation using separable potentials operating only in s-state between different pairs. For convenience, three particles are taken to be identical, spin-less and iso-scalar. The interaction Hamiltonian of three-body system, assuming only pair interactions to operate is taken as

$$V = V_{12} + V_{23} + V_{31}, \quad (2.57)$$

where the matrix-element for V_{ij} is taken as

$$(\vec{P}_i \vec{P}_j | V_{ij} | \vec{P}'_i \vec{P}'_j) = \delta(\vec{p}_k - \vec{p}'_k) (\vec{p}_{ij} | V_{ij} | \vec{p}'_{ij}) \quad (2.58)$$

Assuming the interaction to be separable, acting only in s-state, we write

$$(\vec{p}_{ij} | V_{ij} | \vec{p}'_{ij}) = -\frac{\lambda}{m} g(\vec{p}_{ij}) g(\vec{p}'_{ij}), \quad (2.59)$$

where, as before, we define

$$\vec{p}_{ij} = \frac{\vec{P}_i - \vec{P}_j}{2}, \quad \text{and} \quad \vec{p}_k = \vec{P}_i + \vec{P}_j \quad (2.60)$$

The three-body Schrodinger equation for the wave-function $\psi(\vec{P}_1, \vec{P}_2, \vec{P}_3)$ is expressible [10] as

$$D(E)\psi = - \sum_{1 \neq j \neq k = 1,2,3} m V_{ij} \psi(ij; k),$$

$$D(E) = \frac{1}{2}(P_1^2 + P_2^2 + P_3^2) - mE$$

$$\equiv p_{ij}^2 + \frac{3}{4} p_k^2 - mE \quad (2.61, 2.62)$$

where E is the total energy of the three-body system. In the case of the scattering of a particle by the bound state of the other two

$$E = k^2/2\mu - \alpha^2/m; \quad \mu = \frac{2m}{3}$$

$$= \frac{3}{4}\frac{k^2}{m} - \frac{\alpha^2}{m},$$

where $\vec{k}$ is the incident momentum of the impinging particle and α^2/m is the binding energy of the two-particle state. For the right-hand side of Eq. (2.61), the following operator notation is used:

$$V_{ij}\psi(ij;k) = \int (\vec{P}_i\vec{P}_j|V_{ij}|\vec{P}'_i\,\vec{P}'_j)\,\psi(\vec{P}'_i\,\vec{P}'_j\,\vec{P}_k)\,d\vec{P}'_i\,\vec{P}'_j \tag{2.63}$$

In fact, Eq. (2.63) can be further simplified using Eqs. (2.58) and (2.60). Thus

$$\int (\vec{P}_i\vec{P}_j|V_{ij}|\vec{P}'_i\,\vec{P}'_j\,)\psi(\vec{P}'_i\,\vec{P}'_j\,\vec{P}_k)\,d\vec{P}'_i\,d\vec{P}'_j = -\frac{1}{m}g(\vec{p}_{ij})\,F(\vec{p}_k),$$

$$\text{where} \quad F(\vec{p}_k) = \lambda \int d\vec{p}'_{ij}\,g(\vec{p}'_{ij})\,\psi(\vec{p}'_{ij},\vec{p}_k) \tag{2.64, 2.65}$$

Using Eq. (2.64) on the right-hand side of Eq. (2.61), the structure of the three-body wavefunction can be expressed as [10]

$$D(E)\psi = \psi^{(1)}(\vec{p}_{23},\vec{p}_1) + \psi^{(2)}(\vec{p}_{31},\vec{p}_2) + \psi^{(3)}(\vec{p}_{12},\vec{p}_3)$$

$$= g(\vec{p}_{23})\,F(\vec{p}_1) + g(\vec{p}_{31})\,F(\vec{p}_2) + g(\vec{p}_{12})\,F(\vec{p}_3) \tag{2.66}$$

On using the wavefunction (2.66) on the right-hand side of (2.65), we would get three terms, the first one would be a direct term which is transported to the left while the other two terms can be reduced to have the same structure. We thus finally write

$$[\lambda^{-1} - h(3p_k^2/4 - mE)]\,F(\vec{p}_k) = 2\int d\vec{q}\,\frac{g(\vec{q}+\vec{p}_k/2)\,g(\vec{p}_k+\vec{q}/2)}{q^2 + p_k^2 + \vec{q}\cdot\vec{p}_k - mE - i\varepsilon}F(\vec{q}), \tag{2.67}$$

where

$$h(3p_k^2/4 - mE) = \int d\vec{q}\,\frac{g^2(q)}{(q^2 + 3p_k^2/4 - mE)} \tag{2.68}$$

and λ is the strength parameter of the two-body potential. When the two particles are bound, it is expressed in a close form as

$$\lambda^{-1} = \int d\vec{q}\,\frac{g^2(q)}{(q^2 + \alpha^2)} \tag{2.69}$$

2.7.1 How the Problems of Disconnectedness and the Uniqueness of Boundary Conditions Are Tackled Here?

As a first remark, it is important to observe that the structure of ψ appearing on the right-hand side as a sum of three components $\psi^{(i)}$'s in (2.66) admits of a close resemblance with the corresponding structure in Faddeev's three-particle theory. To make the similarity more transparent, we rewrite

$$\psi(\vec{P}_1, \vec{P}_2, \vec{P}_3) = \psi^{(1)}(\vec{p}_{23}, \vec{p}_1) + \psi^{(2)}(\vec{p}_{31}, \vec{p}_2) + \psi^{(3)}(\vec{p}_{12}, \vec{p}_3), \qquad (2.70)$$

where for the case of three identical particles,

$$\psi^{(1)} = \psi^{(2)} = \psi^{(3)} \quad \text{and} \quad \psi^{(1)} = D^{-1}(E) g(\vec{p}_{23}) F(\vec{p}_1)$$

satisfies the integral equation

$$\psi^{(1)}(\vec{p}_{23}, \vec{p}_1) = \frac{-2m}{p_{23}^2 + 3p_1^2/4 - mE} \int d\vec{q} \, \langle \vec{p}_{23} | t(mE - 3p_1^2/4) | \vec{q} + \vec{P}_1/2 \rangle \psi^{(1)}$$
$$(\vec{p}_1 + \vec{q}/2, \vec{q}) \qquad (2.71)$$

Further, by transporting the direct term to the left in writing the integral Eq. (2.67), we have eliminated in a natural way the problem of disconnectedness of the scattering matrix. The two terms which have been merged together in the case of identical particle having same two-particle scattering amplitude represent the connected part of the three-body kernel. In Eq. (2.71), the two-particle scattering amplitude in s-state is given by

$$\langle \vec{p}_{23} | t(mE - 3p_1^2/4) | \vec{q} + \vec{p}_1/2 \rangle = -\frac{\lambda}{m} \frac{g(\vec{p}_{23}) g(\vec{q} + \vec{p}_1/2)}{[1 - \lambda h(mE - 3p_1^2/4)]} \qquad (2.72)$$

Also note that each term in Eq. (2.66) finds a simple physical interpretation. The term, $D^{-1}(E) g(\vec{p}_{23})$, e.g., can in fact be regarded as a two-body wave function of particles 2 and 3, with the quantity $3p_1^2/4 - mE$, now playing the role of representing the binding energy of the two-body system. The function $F(\vec{p}_1)$, called as a spectator function, where the momentum $\vec{p}_1$ is actually the momentum of the first particle relative to the center of mass of the other two particles represents the wave function of particle 1 in the presence of the other two whose dynamical effects are already contained in the right-hand side of Eq. (2.67). Indeed, Eq. (2.67) can be looked upon as an 'effective two-body Schrodinger equation' describing the behavior of particle 1 with respect to the other two particles which form a bound state. This fact can be appreciated more clearly by noting that

$$[\lambda^{-1} - h(3p_1^2/4 - mE)] = m(p_1^2/2\mu - k^2/2\mu) H(3p_1^2/4 - mE), \qquad (2.73)$$

where

$$H(3p_1^2/4 - mE) = \int d\vec{q}\, \frac{g^2(q)}{(q^2 + \alpha^2)\,(q^2 + 3p_1^2/4 - mE)},$$

which clearly means that the function $F(\vec{p}_1)$ exhibits a pole at $p_1^2 = k^2$. Thus redefining $H(3p_1^2/4 - mE)\,F(\vec{p}_1)$ by $G(\vec{p}_1)$ in Eq. (2.67), we get

$$(p_1^2/2\mu - k^2/2\mu)\,G(\vec{p}_1) = \frac{2}{m} \int d\vec{q}\, \frac{g(\vec{q} + \vec{p}_1/2)\,g(\vec{p}_1 + \vec{q}/2)\,G(\vec{q})}{(q^2 + p_1^2 + \vec{q}.\vec{p}_1 - mE)\,H(3q^2/4 - mE)} \tag{2.74}$$

It follows from this equation that to describe the scattering process, the function $G(\vec{p}_1)$ can be used to assign the proper boundary condition consisting of an incident plane wave plus an outgoing scattered wave, which in momentum space means

$$G(\vec{p}_1) = (2\pi)^3 \delta(\vec{p}_1 - \vec{k}) + \frac{4\pi f_k(\vec{p}_1)}{p_1^2 - k^2 - i\varepsilon} \tag{2.75}$$

Substituting (2.75) in (2.74) gives us the equation for the scattering amplitude

$$(3/4)4\pi\, f_k(\vec{p}_1) = 2(2\pi)^3 N^2 \frac{g(\vec{p}_1 + \vec{k}/2)\,g(\vec{k} + \vec{p}_1/2)}{(p_1^2 + k^2 + \vec{p}_1.\vec{k} - mE)} + 2(4\pi)$$
$$\int d\vec{q}\, \frac{g(\vec{q} + \vec{p}_1/2)\,g(\vec{p}_1 + \vec{q}/2)\,f_k(\vec{q})}{(p_1^2 + q^2 + \vec{p}_1.\vec{q} - mE)\,(H(3q^2/4 - mE))\,(q^2 - k^2 - i\varepsilon)} \tag{2.76}$$

We have thus finally obtained a simple one variable integral equation for an off-shell scattering amplitude to describe the scattering by bound state of identical particles. The energy E is of course understood to have an imaginary term $i\varepsilon$. As regards the uniqueness problem for the boundary condition to obtain the amplitude for the channel of scattering of particle 'i' by the bound 'jk' system, this can be resolved in the Schrodinger equation approach purely on physical considerations. For instance, considering the case of 3 nucleons taking into account the spin, i-spin degrees of freedom, we consider a system consisting of $(2n + 1p)$ corresponding to the i-spin $I = 1/2$ and $I_z = 1/2$, representing the scattering of neutron by deuteron. If we have only s-state interaction between different pairs, we would then have the interaction between singlet and triplet spin states. As a result, we would get two coupled integral equations for the spectator functions $F_1(\vec{p})$ and $F_2(\vec{p})$. While $F_1(\vec{p})$ may represent the scattering of neutron by deuteron ($^3S_1$) state, the function $F_2(\vec{p})$ will correspond to the scattering of neutron by a $d^*(^1S_0)$ state which is a virtual (anti-bound) state. Thus, for n-d scattering only $F_1(\vec{p})$ will be subject to the boundary condition of incoming plane wave plus an outgoing scattered wave for the scattering of neutron of incident momentum k and the function $F_2(\vec{p})$, will have only an outgoing scattered wave.

Chapter 3
Efimov's Universal Three-Body Effect

3.1 Introduction

Efimov [13] discovered a remarkable physical phenomenon where the Hamiltonian of a three-particle quantum system with the resonantly interacting pairs gives rise to an attractive long-range 'effective' three-body force which can produce a large number of energy levels in a three-particle spectrum. The original Efimov's physical argument is based on the replacement of the two-body potential with a boundary condition, which is essentially equivalent to considering a two-body zero-range interaction, and on the introduction of hyper-spherical coordinates. If the resonant condition is satisfied in these coordinates, the problem becomes separable, and in the equation for the hyperradius R, the long-range, attractive effective potential of the form, $V_0(R) = -[|s_0|^2 + 1/4]/R^2$, appears.

In the following subsection, we set up the preliminaries by writing the expression for the scattering amplitude of two particles at low-energy s-wave scattering in terms of the s-wave phase shifts under the condition when the scattering length 'a' is much larger than the range 'r_0'. The s-wave total cross section is then related by unitarity through the optical theorem to the s-wave phase shifts which can produce resonance at large values of scattering length. In the next subsection, a simple derivation to obtain the expression for attractive long-range effective potential for three identical bosons using the proper boundary conditions for the resonant pair interaction is presented. Finally, we highlight the salient features of the Efimov effect.

V. S. Bhasin and I. Mazumdar, *Few Body Dynamics, Efimov Effect and Halo Nuclei*, SpringerBriefs in Physics, https://doi.org/10.1007/978-3-030-56171-0_3

3.2 Relation Between Total Cross Section and Scattering Length for Two Particles at Low Energies

Consider spin-less particles with no internal degrees of freedom interacting through short-range interaction. This means there exists a range r_0 beyond which the relative motion of the particles is almost free. Efimov effect arises when the two-particle interaction is nearly resonant. From scattering theory, we know that at low energies when $k \ll r_0^{-1}$, only the s-wave is scattered and has a nonzero phase shift δ_0, which is given by

$$\tan \delta_0 \approx -ka \tag{3.1}$$

Here, k is the relative wave number between two particles and a is the scattering length. Also, the s-wave scattering amplitude is expressed as:

$$f = \frac{1}{k \cot \delta_0 - ik} \tag{3.2}$$

From Eq. (3.1), it is clear that for two-body interaction to be resonant at low energies, the scattering length has to be much larger than the range r_0. Also, according to optical theorem which essentially arises from the unitarity relation:

$$\text{Im}(f) = \frac{4\pi \sigma_{\text{tot}}}{k} \tag{3.3}$$

we get from Eq. (3.2),

$$\sigma_{\text{tot}} = \frac{4\pi \sin^2 \delta_0}{k^2} \to_{k \to 0} 4\pi a^2 \tag{3.4}$$

It is thus clear that in the limit when $a \to \pm\infty$, the factor $\sin^2 \delta_0$ in Eq. (3.4) approaches its maximum value (called the unitarity limit). Near unitarity, scattering length a (positive or negative) is a single parameter which determines low-energy scattering (i.e., positive energies) as well as the binding energy of a weakly bound state below the two-body breakup threshold (negative energy). This bound state exists only for positive scattering length and its binding energy is close to $\hbar^2/(ma^2)$, where m is the mass of the particles.

3.3 Efimov Effect in Three-Boson System

For three bosons with position coordinates, $\vec{r}_1$, $\vec{r}_2$, $\vec{r}_3$, let us define Jacobi coordinates

$$\vec{r}_{ij} = \vec{r}_j - \vec{r}_i$$

and

$$\vec{\rho}_{ij,k} = \frac{2}{\sqrt{3}}(\vec{r}_k - \frac{\vec{r}_i + \vec{r}_j}{2}), \quad \text{where } (i, j, k = 1, 2, 3) \tag{3.5}$$

After eliminating the center of mass, the time-independent three-body wave function satisfies free Schrodinger equation of energy E in terms of the Jacobi coordinates as:

$$(-\nabla^2_{r_{12}} - \nabla^2_{\rho_{12-3}} - k^2)\Psi = 0, \quad \text{with } E = \frac{\hbar^2 k^2}{m} \tag{3.6}$$

along with Bethe–Peirel's boundary condition for each pair of the particles given by

$$\frac{1}{r\Psi}\frac{\partial}{\partial r}(r\Psi) \to_{r\to 0} \frac{1}{a} \tag{3.7}$$

For the three identical bosons, the totally symmetric wave function Ψ can be expressed in terms of three components

$$\Psi = \chi(\vec{r}_{12}, \vec{\rho}_{12-3}) + \chi(\vec{r}_{23}, \vec{\rho}_{23-1}) + \chi(\vec{r}_{31}, \vec{\rho}_{31-2}) \tag{3.8}$$

where the component χ (known as Faddeev component) satisfies the equation

$$(-\nabla^2_r - \nabla^2_\rho - k^2)\chi(\vec{r}, \vec{\rho}) = 0 \tag{3.9}$$

Note that the coordinates $\vec{r}_{23}, \vec{\rho}_{23-1}, \vec{r}_{31}, \vec{\rho}_{31-2}$ can be related to the coordinates $\vec{r}_{12}, \vec{\rho}_{12-3}$ as:

$$\vec{r}_{23} = -\frac{1}{2}\vec{r}_{12} + \frac{\sqrt{3}}{2}\vec{\rho}_{12-3}, \quad 1 \text{ Equivalently, } \begin{pmatrix} \vec{r}_{23} \\ \vec{\rho}_{23-1} \end{pmatrix} = \begin{pmatrix} -\frac{1}{2} & \frac{\sqrt{3}}{2} \\ -\frac{\sqrt{3}}{2} & -\frac{1}{2} \end{pmatrix}\begin{pmatrix} \vec{r}_{12} \\ \vec{\rho}_{12-3} \end{pmatrix}$$

$$\vec{\rho}_{23-1} = -\frac{\sqrt{3}}{2}\vec{r}_{12} - \frac{1}{2}\vec{\rho}_{12-3},$$

$$\vec{r}_{31} = -\frac{1}{2}\vec{r}_{12} - \frac{\sqrt{3}}{2}\vec{\rho}_{12-3}, \quad \text{or } \begin{pmatrix} \vec{r}_{31} \\ \vec{\rho}_{31-2} \end{pmatrix} = \begin{pmatrix} -\frac{1}{2} & -\frac{\sqrt{3}}{2} \\ \frac{\sqrt{3}}{2} & -\frac{1}{2} \end{pmatrix}\begin{pmatrix} \vec{r}_{12} \\ \vec{\rho}_{12-3} \end{pmatrix}$$

$$\vec{\rho}_{31-2} = \frac{\sqrt{3}}{2}\vec{r}_{12} - \frac{1}{2}\vec{\rho}_{12-3} \tag{3.10}$$

Now applying Bethe–Peirel's boundary condition to Eq. (3.8) for the pair (1, 2), we get

$$\left[\frac{\partial}{\partial r}(r\chi(\vec{r},\vec{\rho}))\right]_{r\to 0} + \chi\left(\frac{\sqrt{3}}{2}\vec{\rho}, -\frac{1}{2}\vec{\rho}\right) + \chi\left(-\frac{\sqrt{3}}{2}\vec{\rho}, -\frac{1}{2}\vec{\rho}\right)$$

$$= \left[-\frac{r}{a}\left\{\chi(\vec{r},\vec{\rho}) + \chi\left(\frac{\sqrt{3}}{2}\vec{\rho}, -\frac{1}{2}\vec{\rho}\right) + \chi\left(-\frac{\sqrt{3}}{2}\vec{\rho}, -\frac{1}{2}\vec{\rho}\right)\right\}\right]_{r\to 0}$$

$$\tag{3.11}$$

In Eq. (3.11), the vectors $\vec{r} \equiv \vec{r}_{12}$ *and* $\vec{\rho} \equiv \vec{r}_{12-3}$. In the limit $r \to 0$, only the first term on the right-hand side survives because $\chi(\vec{r},\vec{\rho})$ diverges when $r \to 0$ but is finite elsewhere. To simplify further, we expand χ in partial waves and restrict only to the case when total angular momentum $L = 0$, where we write

$$\chi(\vec{r},\vec{\rho}) = \frac{\chi_0(r,\rho)}{r\rho} \tag{3.12}$$

$\chi_0(r,\rho)$ is finite for $r \to 0$, but it must also satisfy the condition

$$\chi_0(r,\rho) \to_{\rho\to\infty} 0 \tag{3.13}$$

to ensure that χ_0 remains finite in this limit. We now substitute (3.12) into Eq. (3.9) and Eq. (3.11) to get

$$\left[-\frac{\partial^2}{\partial r^2} - \frac{\partial^2}{\partial \rho^2} - k^2\right]\chi_0(r,\rho) = 0 \tag{3.14}$$

and the boundary condition for $r \to 0$

$$\left[\frac{\partial}{\partial r}(\chi_0(r,\rho))\right]_{r\to 0} + 2\frac{1}{\frac{\sqrt{3}}{4}\rho}\chi_0\left(\frac{\sqrt{3}}{2}\rho, \frac{1}{2}\rho\right) = -\frac{1}{a}\chi_0(0,\rho) \tag{3.15}$$

To proceed further, we transform the coordinates (r, ρ) into hyperspherical coordinates [24] (polar coordinates) by defining:

$$r = R \sin\alpha,$$
$$\rho = R \cos\alpha, \tag{3.16}$$

where R is the hyper radius given by

$$R^2 = r^2 + \rho^2 = \frac{2}{3}\left(r_{12}^2 + r_{23}^2 + r_{31}^2\right) \tag{3.17}$$

and α is the Delves hyper-angle:

$$\tan \alpha = \frac{r}{\rho} \tag{3.18}$$

In terms of these coordinates, one has the equation:

$$\left[-\frac{\partial^2}{\partial R^2} - \frac{1}{R}\frac{\partial}{\partial R} - \frac{1}{R^2}\frac{\partial^2}{\partial \alpha^2} - k^2 \right] \chi_0(R, \alpha) = 0 \tag{3.19}$$

with the boundary condition:

$$\left[\frac{\partial}{\partial \alpha}(\chi_0(R, \alpha)) \right]_{\alpha \to 0} + \frac{8}{\sqrt{3}} \chi_0(R, \pi/3) = -\frac{R}{a} \chi_0(R, 0) \tag{3.20}$$

In the unitarity limit when $a \to \pm\infty$, the right-hand side of Eq. (3.20) becomes zero. The equation then gets separable in R and α. In this limit, one is left with the boundary condition at $\alpha = 0$, which is independent of R. For the other boundary condition Eq. (3.13) corresponding to the case when $\chi_0(R, \pi/2) = 0$ at $\alpha = \pi/2$, the equation is again independent of R. One thus finds a solution of Eq. (3.19) expressing:

$$\chi_0(R, \alpha) = G(R)\, \Phi(\alpha) \tag{3.21}$$

where $\Phi(\alpha)$ satisfies the equation:

$$-\frac{\partial^2}{\partial \alpha^2}\Phi(\alpha) = s_n^2 \Phi(\alpha) \tag{3.22}$$

with the boundary conditions at $\alpha = 0$ *and* $\alpha = \pi/2$. This gives the solutions:

$$\Phi_n(\alpha) = \sin(s_n(\pi/2 - \alpha)) \tag{3.23}$$

where s_n is a solution of the equation:

$$-s_n \cos(s_n \pi/2) + \frac{8}{\sqrt{3}} \sin(s_n \pi/6) = 0 \tag{3.24}$$

For each solution s_n, there is a corresponding solution of hyper-radial function $G_n(R)$ such that $G_n(R)\, \Phi_n(\alpha)$ is a solution of Eq. (3.19). The radial function $G_n(R)$ satisfies the equation:

$$\left(-\frac{\partial^2}{\partial R^2} - \frac{1}{R}\frac{\partial}{\partial R} + s_n^2 - k^2 \right) G_n(R) = 0$$

or equivalently

$$\left(-\frac{\partial^2}{\partial R^2} + V_n(R) - k^2 \right) \sqrt{R} G_n(R) = 0 \tag{3.25}$$

All the solutions of transcendental Eq. (3.24) are real except the one for which $s_0 \approx \pm 1.00624\, i$, which is purely imaginary. As a result, in the case of resonantly interacting particles, the channel $n = 0$ leads to an effective three-body attraction, given by

$$V_0(R) = -\frac{|s_0^2| + 1/4}{R^2} \tag{3.26}$$

This result forms the basis of the Efimov physics.

3.4   Salient Feature of Efimov Effect

Fundamentally important findings of Efimov's investigations can thus be summarized as:

Efimov has shown that there exists an effective long-range attractive interaction of kinetic origin produced by the resonance condition and is independent of the details of the two-body potentials. Physically, this attractive interaction may be interpreted as a mediated attraction between the two particles by the exchange of the third one which is moving back and forth between the two. An important consequence of the effective three-body potential $1/R^2$ is that as long as $|a|$ is sufficiently large compared to the range r_0, there is a sequence of three body bound states whose binding energies form a geometric series lying in the interval between $\hbar^2/mr_0^2$ and $\hbar^2/ma^2$. As $|a|$ is increased, new bound states appear in the spectrum at critical values of a determined by a multiplicative factor $\exp(\pi/s_0)$, where in the case of identical bosons, s_0 is the solution of the transcendental equation:

$$s_0 \cosh \frac{\pi\, s_0}{2} = \frac{8}{\sqrt{3}} \sinh \frac{\pi\, s_0}{6} \tag{3.27}$$

[Compare this equation with Eq. (3.24), where s_0 is taken as imaginary.] The numerical value of s_0 is ≈ 1.00624 or $\exp(\pi/s_0) \approx 22.7$. The number of bound states in the asymptotic limit, as $|a|/r_0 \rightarrow \infty$, is

$$N \rightarrow \frac{s_0}{\pi} \ln(|a|/r_0) \tag{3.28}$$

Figure 3.1 is a very nice representation of the underlying physics of the Efimov effect and is guided by similar representations in the original works of Efimov [13]. There are infinitely many three-body bound states with an accumulation point at the three-body threshold in the limit, and in the resonant limit, the ratio of the binding energies of the successive bound states approaches a universal number near threshold, i.e.,

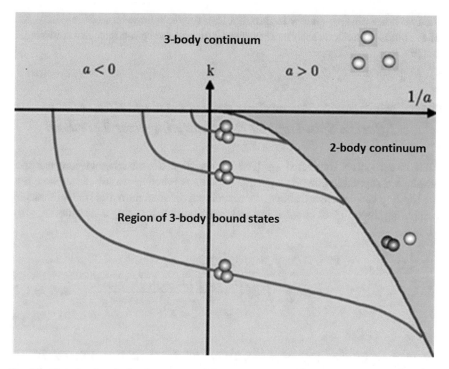

Fig. 3.1 Plot showing the level spectrum of three neutral spinless particles in the $(k$-$a^{-1})$ plane, where k is the energy variable for each Efimov state and a is the scattering length

$$E_T^{(n+1)}/E_T^{(n)} \to \exp(-2\pi/s_0) \quad \text{as } n \to \infty \quad at \ a = \pm\infty \qquad (3.29)$$

Suffice it to emphasize that Efimov effect describes a universal low-energy behavior of the three-particle system. As a consequence of this universality character, the effect can be realized and observed in various physical contexts, e.g., in atomic, molecular, nuclear or condensed matter physics. Efimov has also shown in subsequent papers [25] that this scaling limit does not only hold in the unitarity limit $a \to \pm\infty$ but also occurs for large scattering length.

3.5   Predictions on the Properties of Three-Nucleon System

Efimov, in one of the papers [26] studying the universal low-energy properties in three-nucleon bound and scattering states, conjectured the possibility of 'anomalous behavior' in the n-d scattering cross section near three-body threshold. Following this, Tracheko [27] studied the behavior of $k \cot \delta$ in $^2S_{1/2}$ n-d scattering, employing a three-body model which involves the effective range of approximation for the two-body t-matrix, and found such an anomalous behavior does exist not only when the two-body binding energy is zero, but also for realistic values of the parameters of the nucleon–nucleon amplitude corresponding to the spin-triplet and spin-singlet s-states.

3.5.1   Studying the Anomalous Behavior Due to Efimov Effect in Spin-Doublet N-D Scattering Near Threshold

This investigation motivated us [28] to revisit n-d scattering at low energies employing separable potentials. In order to check whether we are close enough to the limiting case discussed above, we carried out detailed numerical calculations by computing the off-shell coupled integral equations for the $^2S_{1/2}$ n-d scattering [10], viz.

$$4\pi \, k_\alpha(p, k) \, f_\alpha(p, k) = -(2\pi)^3 D^{-1}(\vec{p}, \vec{k}) \, K_{\alpha 1}(\vec{p}, \vec{k})$$
$$+ 4\pi \sum_\beta \int d\vec{q} \, \frac{D^{-1}(\vec{p}, \vec{q}) \, K_{\alpha\beta}(\vec{p}, \vec{q}) \, f_\beta(\vec{q}, k)}{q^2 - k^2 - i\varepsilon}, \quad \alpha, \beta = 1, 2$$

(3.30)

where $k_\alpha(p.k) = [\lambda_\alpha^{-1} - h_\alpha(p, k)] \, (p^2 - k^2)^{-1}$ and $[K_{\alpha\beta}(\vec{p}, \vec{q})] = \begin{pmatrix} \frac{1}{2} g(\vec{\xi}) \, g(\vec{\eta}) & -\frac{3}{2} g(\vec{\xi}) \, f(\vec{\eta}) \\ -\frac{3}{2} f(\vec{\xi}) \, g(\vec{\eta}) & \frac{1}{2} f(\vec{\xi}) \, g(\vec{\eta}) \end{pmatrix}$ with $\vec{\xi} = \vec{q} + \frac{1}{2}\vec{p}; \vec{\eta} = \vec{p} + \frac{1}{2}\vec{q};$ and $h_1(p) = \int d\vec{q} \, \frac{g^2(\vec{q})}{q^2 + 3p^2/4 + \alpha^2 - 3k^2/4 - i\varepsilon}, \quad h_2(p) = \int d\vec{q} \, \frac{f^2(q)}{q^2 + 3p^2/4 + \alpha^2 - 3k^2/4 - i\varepsilon}$

Here, λ_1 refers to the spin-triplet and isospin-singlet while λ_2 represents the spin-singlet, isospin-triplet strength parameters. The function $f_1(p, k)|_{p=k}$ is the s-wave scattering amplitude for n-d scattering, while $f_2(p, k)$ is the auxiliary function representing n-d^* system, where d^* is the virtual deuteron state. Detailed numerical calculations are given in [28]. The plot of $k \cot \delta$ versus k^2 clearly demonstrates that in the physical case, we do not see any anomaly in the behavior of $k \cot \delta$, whereas in cases where we consider the values of $a_s = -100$ fm or $-\infty$, there does appear a dip structure which is sharpened as we increase the value of singlet scattering length $|a_s|$.

It is worthwhile pointing out that just after Efimov's paper appeared, Amado and Noble [29] studying the analytic properties of the Fredholm determinant in a three-boson model showed that with increase in potential strength, the Efimov states move into the unphysical sheet associated with the unitarity cut. A detailed analysis was followed by Adhikari et al. [30], to study the movement of the Efimov states in the three-boson model and in the s-wave spin-doublet ($^2S_{1/2}$) three-nucleon system. The key question addressed there was whether the Efimov states, with the increase in potential strength, move over to the virtual states in the unphysical sheet or two of the Efimov states collide to produce a resonance pair, one of which may come close to the scattering region and produce an observable effect on the physical scattering process.

In the Sect. 6 on halo nuclei, we encountered precisely such a situation where we found two Efimov states in the case of halo nucleus ^{20}C, and while studying $n - {}^{19}$C scattering, we found these Efimov states moved over to produce a resonance just near threshold. We shall discuss these details in Chap. 6.

3.5.2 Appearance of Efimov Effect in Three-Body Separable Potential Approach

To conclude this study, it would be interesting to record how the effect of long-range three-body potential of the type $1/R^2$ as originally found by Efimov appears in the three-body problem with separable potentials. It should be noted that to analyze the Efimov effect in the limit when two-body scattering length is large as compared to the short-range potential (corresponding to large values of β in the present case), the kernel of the three-body Schrodinger equation is basically governed by the function $H(p)$, defined by Eq. (2.73), where

$$H(p) = [\lambda^{-1} - h(p)]^{-1} = \left[\frac{\pi^2}{\beta(\beta + \alpha)^2} - \frac{\pi^2}{\beta[\beta + \sqrt{\alpha^2 - 3k^2/4 + 3p^2/4}]^2} \right]^{-1}$$

So as long as the two-body potential is sufficiently strong to produce a three-body bound state with $k^2 < 0$, $H(p)$ will be a regular function of p for real p. However, as $k^2 \to 0$, three-body binding energy approaches that of the two-particle system represented by the parameter α^2. In such a case,

$$H(p) = \left[\frac{\pi^2}{\beta(\beta + \alpha)^2} - \frac{\pi^2}{\beta[\beta + \sqrt{\alpha^2 + 3p^2/4}]^2} \right]^{-1} \Rightarrow_{p \to 0} \frac{4\alpha\beta(\alpha + \beta)^3}{3\pi^2 p^2}$$

Thus, $H(p)$ becomes singular near $p = 0$. Further, in the limiting case, where in addition $\alpha = 0$, corresponding to infinite scattering length, we find

$$H(p) = \left[\frac{\pi^2}{\beta^3} - \frac{\pi^2}{\beta[\beta + \sqrt{3}p/2]^2} \right]^{-1} \Rightarrow_{p \to 0} \frac{\beta^4}{\sqrt{3}\pi^2 p}.$$

This $1/p$ singular behavior in momentum space corresponds to the singularity of the type $1/R^2$ in configuration space leading to the Efimov effect [31]. To see that for $\alpha = 0$, the scattering length does become infinite, we have

$$H(p) = \lambda[1 - \lambda h(p)]^{-1} \Rightarrow_{p \to 0} \lambda[1 - \lambda\frac{\pi^2}{\beta^3}]^{-1}$$

On comparing this with the expression for the scattering length in this model, we find

$$a = -\frac{2\pi^2\lambda}{\beta^4}\left[1 - \frac{\lambda\pi^2}{\beta^3}\right]^{-1}$$

In order to have infinite scattering length, $\lambda\pi^2/\beta^3 = 1$, which is precisely the value that we get for $\alpha = 0$.

This completes our discussion on the investigations taken up immediately after Efimov's observation of the existence of an effective long-range attractive three-body force and its possible impact on the physical phenomena observed in three-body systems. The detailed systematic studies carried out in different areas in nuclear physics demonstrating the experimental evidence of Efimov effect will be discussed in the subsequent chapter.

Chapter 4
Effective Field Theory

4.1 Introduction

Effective field theory offers a general model independent approach to study the low energy behavior of physical systems in several different branches of physics; most importantly in quantum electrodynamics, standard model of elementary particle physics and condensed matter physics. While most of the applications of effective field theories are carried out within the framework of quantum field theory, it has been demonstrated by Lapage [32] that the general principles of effective theory can equally well be applied to non-relativistic quantum mechanics for two and many particle systems within the framework of potential approach.

In this chapter, we present a brief description of low energy N-N scattering and three-body scattering problem with a view to illustrate the basic principles of effective field theory, including the renormalization procedure employed to calculate the loop diagrams involved in the scattering amplitude. We then derive, starting from a three-body Schrodinger equation using two-body s-state separable potentials, the expressions for the scattering amplitude in the framework of effective field theory to study n–^{19}C scattering at low energies (where ^{19}C is assumed to be a loosely bound state of n–^{18}C system).

4.2 Two-Body Problem: N-N Scattering at Low Energies in Effective Field Theory

A general principle of effective field theory, used in quantum mechanical problems employing two body potentials, is that if we are interested only in low energy observables where the large distance behavior of the potential $V(r)$ is dominant, then such an interaction can be replaced by an effective potential, $V_{\text{eff}}(r; \lambda)$, where V_{eff} is supposed

V. S. Bhasin and I. Mazumdar, *Few Body Dynamics, Efimov Effect and Halo Nuclei*, SpringerBriefs in Physics, https://doi.org/10.1007/978-3-030-56171-0_4

Fig. 4.1 Feynman diagram:
two body contact interaction

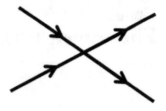

to describe the interaction for $r > r_0$ and whose behavior at short distances $r < r_0$ can be replaced by something simpler with the help of the adjustable parameter λ.

The simplest text-book example [33] can be considered from the study of scattering of sufficiently low thermal neutrons (when $\lambda \gg r_0$) by molecular hydrogen where the effective potential between neutron and proton is assumed in the form of a delta-function, viz;

$$V_{\text{eff}}(r) = (2\pi\hbar^2/\mu)a\delta(\vec{r}), \qquad \vec{r} = \vec{r}_n - \vec{r}_p,$$

where $\vec{r}_n$ and $\vec{r}_p$ are the position vectors of neutron and proton respectively and the coefficient of the delta function (called contact potential) is obtained from the requirement that the scattering amplitude in the first-order perturbation theory coincides with the scattering length, i.e., $f = -a$; here (μ) is the reduced mass of neutron and proton. Thus, by studying elastic scattering of slow neutrons in ortho- and para-hydrogen and comparing the total cross sections, σ_{para} and σ_{ortho}, with the experimental data, it is possible to determine the absolute values of a_t and a_s as well as the sign of the ratio a_t/a_s.

In the language of effective field theory, this two-body contact interaction, as shown by the Feynman diagram (Fig. 4.1) can be described by writing the Lagrangian density as

$$L = \psi^+\left(i\frac{\partial}{\partial t} + \frac{1}{2m}\nabla^2\right)\psi - \frac{C_0}{4}(\psi^+\psi)^2 \qquad (4.1)$$

In formulating the concept of effective field theory, an important ingredient used is the decoupling principle. We know that the theory of strong interaction at distances small compared to $h/(M_{\text{QCD}}c)$, where $M_{\text{QCD}} \approx 1\text{GeV}/c^2$ is the hadron mass scale, is described by Quantum Chromodynamics (QCD). At large distances, QCD is non-perturbative in its coupling constant and is more easily described in terms of chiral quantum field theory at distances comparable to $r_0 \approx h/(m_\pi c)$. For distances much larger than r_0, pion exchange can be considered as a short-range effect and nuclear interaction reduces to delta function and their derivatives. This contact, or pionless effective field theory becomes relevant for light nuclei where the N-N scattering length is much larger than r_0.

To extend the study for s-wave nucleon-nucleon scattering, we calculate the scattering amplitude as a power series in C_0 using perturbation theory. The first three

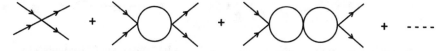

Fig. 4.2 Perturbative expansion of leading order Feynman diagrams in powers of C_0

Feynman diagrams in the perturbation expansion of the scattering amplitude are shown here in Fig. 4.2.

In the low energy domain of two-nucleon system, where the scattering length $|a| \gg |r_{\text{eff}}| \approx r_0$ and the contact or pionless effective field theory [34, 35] can be used without including the contributions of derivative terms, the two nucleon s-wave scattering amplitude in leading order becomes particularly simple.

In general, the amplitude depends on the energies and momenta of the four external lines. This is called an off-shell amplitude. Energy and momentum are conserved at every vertex of the Feynman diagrams. For every closed loop, the energy and momentum are to be integrated. In the center of mass frame, let $\vec{k}$ and $-\vec{k}$ be the two incoming momenta and $\vec{p}$ and $-\vec{p}$ represent outgoing momenta of the scattered nucleons. The amplitude then depends on $\vec{k}$, $\vec{p}$ and the off-shell energies. In the presence of two-body contact interaction, the amplitude gets simplified considerably. In the center of mass frame, the on-shell amplitude is a function of the energy $E = k^2/2\mu = p^2/2\mu$, where $\mu = m/2$.

In order to illustrate the renormalization procedure, let us calculate the scattering amplitude by concentrating only the two simplest diagrams; the single and once iterated contact interactions each in s-wave channel:

$$T_{1c} = -C_{0s}$$

$$T_{2c} = -iC_{0s}^2 \int \frac{d^3q}{(2\pi)^3} \int \frac{dq_0}{(2\pi)} \frac{1}{q_0 - q^2/2m + i\varepsilon} \frac{1}{E - q_0 - q^2/2m + i\varepsilon}$$

$$= -C_{0s}^2 \int \frac{d^3q}{(2\pi)^3} \frac{1}{E - \frac{q^2}{m} + i\varepsilon} \tag{4.2}$$

The last integral over $\vec{q}$ diverges. It has to be regularized by imposing an ultraviolet cutoff Λ so that the integral can be simplified as

$$C_{0s}^2 \int \frac{d^3q}{(2\pi)^3} \frac{1}{E - \frac{q^2}{m} - i\varepsilon} = \frac{C_{0s}^2 m}{(2\pi^2)} \int_0^\Lambda \frac{q^2 dq}{-(q^2 + \alpha^2)}$$

$$= -\frac{C_{0s}^2 m}{2\pi^2} \left[q - \alpha \tan^{-1} \frac{q}{\alpha} \right]_0^\Lambda \xrightarrow{\Lambda \gg \sqrt{E}} -\frac{C_{0s}^2 m}{2\pi^2} \left(\Lambda - \frac{\pi}{2} \alpha \right)$$

where the constant $\alpha = \sqrt{-mE - i\varepsilon}$. The net scattering amplitude thus reduces to

$$T = T_{1c} + T_{2c} = -C_{0s} + \frac{C_{0s}^2 m}{2\pi^2}\left(\Lambda - \frac{\pi}{2}\sqrt{-mE - i\varepsilon}\right) \qquad (4.3)$$

The dependence on the ultraviolet cutoff Λ can be consistently eliminated by a perturbative renormalization procedure [22]. A simple possibility is to identify the constant terms (independent of k) with the scattering length a in terms of the coupling parameter, which is

$$a = \frac{m}{4\pi}C_{0s}\left(1 - \frac{C_{0s}m\Lambda}{2\pi^2} + \ldots\ldots\right) \qquad (4.4)$$

Alternatively,

$$a = \frac{m}{4\pi}C_{0s}\left(1 + \frac{C_{0s}m\Lambda}{2\pi^2}\right)^{-1} \qquad (4.5)$$

Inverting this expression to obtain $1/a$ as a function of C_{0s}, we get

$$\frac{1}{a} = \frac{4\pi}{mC_{0s}}\left(1 - \frac{C_{0s}m\Lambda}{2\pi^2}\right)^{-1} = \left(\frac{4\pi}{mC_{0s}} + \frac{2\Lambda}{\pi}\right) \qquad (4.6)$$

or equivalently,

$$C_{0s} = \frac{4\pi a}{m}\left(1 - \frac{2\Lambda a}{\pi}\right)^{-1} \qquad (4.7)$$

Finally, the s-wave scattering amplitude is written as

$$T^{(0)} = \frac{4\pi}{m}\left[-\left(\frac{4\pi}{mC_{0s}} + \frac{2\Lambda}{\pi}\right) - ik\right]^{-1}\left[1 + O\left(\frac{k}{\Lambda}\right)\right] \qquad (4.8)$$

to be compared with the standard expression

$$T^{(0)} = \frac{4\pi}{m}\left[-\frac{1}{a} + \frac{1}{2}r_0 k^2 + \ldots\ldots - ik\right]^{-1} \qquad (4.9)$$

The comparison of the expression Eq. (4.8) with Eq. (4.9) shows how the amplitude derived in effective field theory recovers in the lowest order with the standard expression in effective range theory by defining the renormalized coupling

$$C_{0s}^R = C_{0s}\left[1 + \frac{mC_{0s}\Lambda}{2\pi^2}\right]^{-1} = C_{0s}\left(1 - \frac{C_{0s}m\Lambda}{2\pi^2}\right) = \frac{4\pi a}{m} \qquad (4.10)$$

The expression for the amplitude shows a pole at $k = i\kappa = i/a$ which corresponds to a real or virtual bound state depending on whether the scattering length has a positive or negative sign with binding energy, BE $= 1/(ma^2)$.

From the point of view of two nucleon systems at low energies, it is worth recalling that the binding energy BE of deuteron is about 2.2 MeV, which corresponds to the deuteron binding momentum, $\aleph \approx 45$ MeV, which is smaller than pion mass by a factor of about 3 indicating that the two nucleons are effectively at a distance three times larger than the range of the nuclear force. In the case of spin singlet state $^1S_0$, the situation is even more dramatic: with negative BE ≈ -0.07 MeV which corresponds to the momentum, $\aleph \approx 8$ MeV, which is 20 times smaller than the pion mass. Thus, deuteron as well as the virtual spin singlet s-state characterized by large negative scattering length prove to be the realistic simplest examples of the weakly bound two-body systems which can be studied by effective field theory using contact interaction.

From the expression (4.5) for the scattering length, it becomes clear that the scattering length diverges when C_{0s} is tuned to the value $-2\pi^2/(m\Lambda)$. This behavior is peculiar and illustrates an important basic principle of effective theories. Normally, non-analytic behavior in long-distance observables arises from the analytic structure at short distance parameters. Here the short-distance parameter is the strength of the two-body contact interaction while the long-distance observable is the scattering length. This only shows that divergence in a is essentially arising due to the quantum fluctuations involving virtual particles with wave numbers less than Λ where the particles with wave numbers greater than Λ are excluded by this cutoff. It should also be noticed from the expression (4.7) that the coupling parameter C_{0s} diverges as $\Lambda \to \pi/(2a)$. The physical observables are, however, independent of the parameter, Λ. It is therefore obvious that the effects of arbitrary large coupling constant have to be compensated by equally large effects of quantum fluctuations involving virtual particles.

From the point of view of the renormalization group, we start renormalizing the expression (4.7) by multiplying both sides by $m\Lambda/(2\pi^2)$ and redefining the dimensionless coupling constant as $\widehat{C}_{0s} \equiv \frac{m\Lambda C_{0s}}{2\pi^2}$. Thus, we re-express (4.7) as

$$\widehat{C}_{0s} = \frac{2\Lambda a/\pi}{1 - 2\Lambda a/\pi} = -\frac{a\Lambda}{a\Lambda - \pi/2} \qquad (4.11)$$

in terms of the coupling constant $\widehat{C}_{0s}$. We note that as Λ is varied keeping, a, fixed, the plot of $\widehat{C}_{0s}$ as a function of Λ maps a trajectory, called a renormalization group (RG) trajectory. As Λ is increased, approaching infinity, the constant $\widehat{C}_{0s}$ in Eq. (4.11) approaches an ultraviolet fixed point, i.e.,

$$\widehat{C}_{0s}(\Lambda) \to -1 \quad \text{as} \quad \Lambda \to \infty \qquad (4.12)$$

Alternatively, we may express the renormalization group as a differential equation

$$\Lambda \frac{\mathrm{d}}{\mathrm{d}\Lambda}\widehat{C}_{0s} = -\frac{a\Lambda}{a\Lambda - \pi/2} + \frac{a^2\Lambda^2}{(a\Lambda - \pi/2)^2} = \widehat{C}_{0s}(1 + \widehat{C}_{0s}) \qquad (4.13)$$

From this equation, it is clear that the RG flow of $\widehat{C}_{0s}$ has two fixed points: $\widehat{C}_{0s} = -1$ and $\widehat{C}_{0s} = 0$. The fixed point $\widehat{C}_{0s} = -1$ corresponds to the resonant limit $a \to \pm\infty$ [see Eq. (4.5)] while the fixed point $\widehat{C}_{0s} = 0$ represents the non-interacting system with $a = 0$. The perturbative expansion of the scattering amplitude (cf. Eq. (4.3) along with (4.4) corresponds to an expansion about this fixed point.

To sum up, effective field theory is defined by a Lagrangian containing all possible local terms consistent with the symmetries of the underlying theory. Since this leads to an infinite number of terms, the goal is to arrange these terms according to some power counting scheme depending on the low-energy scales in each term. The renormalization group (RG) provides us a mathematical tool to determine this power counting. The ultraviolet divergences from high momentum modes of loop diagrams in the Lagrangian can be regulated by a floating ultraviolet cutoff Λ. As all the observables are to be independent of the cutoff, the theory is normalized by absorbing the Λ-dependence into the effective field theory couplings. Thus, the framework for constructing power counting schemes is provided by the resulting RG differential equation.

4.3   Three-Body Scattering Problem in Effective Field Theory

In the three-body sector, one of the earliest attempts using pair-wise contact (or zero range) interaction in three-body problem was made by G. Skorniakov and K. Ter-Marterosian [36] to derive the integral equation studying three-body bound state and scattering of neutrons by deuterons at low energies. It was soon realized that in the case of zero-range interaction, the resulting scattering amplitude for spin $S = 3/2$ (quartet total angular momentum) channel yields a unique solution, whereas for the spin doublet ($S = 1/2$) channel the corresponding coupled integral equations did not yield a unique result. This could be easily explained on the basis of the fact that in the quartet channel, the effective interaction between neutron and deuteron happens to be repulsive due to Pauli principle and as a result the zero-range force becomes ineffective. On the other hand, in the doublet channel, the effective interaction is attractive and therefore the finite range effects of the interaction which play an important dynamical role can no longer be neglected here. A practical solution to this problem was suggested by V.N. Gribov and demonstrated by G. Danilov [37], which consisted in imposing a condition on the solutions of three-body equation to reproduce the known three-body observable. Thus, for instance, fixing the triton binding energy to the observed value and solving the three-body equation it is possible to predict the n-d doublet scattering length. Indeed, this correlation between the two basic quantities, viz., n-d scattering length and triton binding energy attracted a great deal of attention in recent years since the time it was pointed out by Phillips [37(b)].

Within the framework of field theory using Feynman diagrams, STM equations were first derived by V. Komarov and A. Popova [38], in investigating the breakup of deuterons by neutrons at low energies. Studying the scattering of a third particle by the bound state pair, the two-body construct potential leads to a particle exchange force with a range of the order of two-body scattering length. For instance, in n-d elastic scattering the lowest order Born term arises from the exchange of proton by deuteron, which in the language of nuclear physics is known as a pick-up process. Thus, in nuclear physics, the length scale set by the scattering length is far larger than any other scale. This makes it worthwhile using this separation of scales to construct effective field theory for three-body scattering systems.

In the seminal work of Bedaque and collaborators [22], the derivation of STM equation is found to be particularly simple in the effective field theory, realizing the fact that STM integral equation is not an equation for the six-point Green's function but is an equation for the four-point Green's function, $\langle 0|T(d\psi d^*\psi^*)|0\rangle$, where the two-particle composite dimer field operator d is annihilating the two particles. In order to obtain a unique solution, Bedaque et al. [22] also introduced a three-body force in the leading order three-body EFT equation and adjusted this force to reproduce the experimental $1 + 2$ scattering length. Earlier, in 1995, Adhikari, Frederico and Goldman [39] also pointed out the need of introducing a three-body observable to account for the divergence appearing in the kernel of the Faddeev equations for zero range interaction.

In an approach alternative to introducing a three-body force, I Afnan and D Phillips [40] revisit the Amado model for the case of three spin-less bosons to study the scattering in which the interaction of an incident particle on a composite system is considered using the framework of Lee model Lagrangian

$$L = N^+\left(i\partial_0 + \frac{\nabla^2}{2M}\right)N + D^+\Delta D + g[D^+NN + DN^+N^+] \qquad (4.14)$$

using a field theory of bosons N in which the two-boson bound state, 'dimer', D is considered explicitly. To obtain a finite amplitude for boson-dimer scattering, the coupling parameter g is replaced by the function $g(p)$, where p is the relative momentum of the two bosons $D \to NN$-a form factor in the Lagrangian. Incidentally, the resulting equation turns out to be essentially the same as obtained from Faddeev's three-body scattering equation using two-body separable potential, [see, e.g., Eq. (2.76)]. The off-shell two-body NN amplitude

$$t(p, p'; E) = g(p)\tau(E)g(p') \qquad (4.15)$$

used here has also the same structure as obtained from the two-body separable potential in s-state. Having performed the renormalization procedure to get the physical value of the binding energy of the dimer state, the cutoff function (or form factor)

$g(p)$ is taken as 1 in the limit so as to reduce the integral equation for the boson-dimer scattering amplitude exactly that obtained in the effective field theory with short-range interaction alone. The main thrust of this approach is to demonstrate, through a detailed analysis studying n-d scattering, that the presence of eigenvalues in the attractive channel in three-body scattering using effective field theory can be removed by one subtraction from the kernel to obtain a renormalization group invariant scattering amplitude.

As we are basically interested in investigating the structural properties of halo nuclei within the framework of three-body system using separable potential, the above approach by Afnan and Phillips can be easily tuned particularly for studying the resonant structure of n–^{19}C(n–dimer(n–^{18}C)) scattering at low energies, using effective field theory.

4.4 Derivation of Three-Body Scattering Amplitude in Effective Field Theory Approaching from Separable Potentials

To derive the equation for the scattering amplitude in effective field theory from s-state separable potential in the three-boson system, we start from Eq. (2.74) for the spectator function $G(\vec{p})$ of third particle in the presence of the other two bound particles.

$$(p_1^2/2\mu - k^2/2\mu)G(\vec{p}_1) = \frac{2}{m} \int d\vec{q} \, \frac{g(\vec{q} + \vec{p}_1/2)g(\vec{p}_1 + \vec{q}/2)G(\vec{q})}{(q^2 + p_1^2 + \vec{q}.\vec{p}_1 - mE)H(3q^2/4 - mE)},$$
(4.16)

where

$$H(3p^2/4 - mE) = \int d\vec{q} \, \frac{g^2(q)}{(q^2 + \alpha^2)(q^2 + 3p^2/4 - mE)}$$
(4.17)

Substituting (4.17) and using the boundary condition

$$G(\vec{p}) = (2\pi)^3\delta(\vec{p} - \vec{k}) + \frac{4\pi f_k(\vec{p})}{p^2 - k^2 - i\varepsilon}$$
(4.18)

we get the integral equation for the scattering amplitude, $f_k(\vec{p})$,

$$(3/4)4\pi f_k(\vec{p}_1) = 2(2\pi)^3 N^2 \frac{g(\vec{p}_1 + \vec{k}/2)g(\vec{k} + \vec{p}_1/2)}{(p_1^2 + k^2 + \vec{p}_1.\vec{k} - mE)} + 2(4\pi)$$

$$\int d\vec{q} \frac{g(\vec{q} + \vec{p}_1/2)g(\vec{p}_1 + \vec{q}/2)f_k(\vec{q})}{(p_1^2 + q^2 + \vec{p}_1.\vec{q} - mE)(H(3q^2/4 - mE))(q^2 - k^2 - i\varepsilon)}$$

$$(4.19)$$

At this stage, in the limit when $g(p)$ is taken as 1, the function $H(3p_1^2 - mE)$ can be reduced as the residue of the pole term of the two-particle scattering amplitude in the three-body Hilbert space, which is simply worked out to be $\frac{2\pi^2}{[\alpha + \sqrt{3p^2/4 - mE}]}$. Similarly, in the zero-range limit, the normalization constant, N, for the bound dimer state is found to be $\sqrt{\alpha}/\pi$, where α is the binding energy parameter of the dimer state. Substituting these expressions in Eq. (4.19) results in the off-energy-shell scattering amplitude for boson-dimer scattering as:

$$\frac{3}{4}(4\pi)f(\vec{p}, \vec{k}; E) = 2(2\pi)^3 \frac{\alpha}{\pi^2} \frac{1}{(\vec{p}^2 + \vec{k}^2 + \vec{p}.\vec{k} - mE)}$$

$$+ 2(4\pi)/(2\pi^2) \int d\vec{q} \frac{(\alpha + \sqrt{3q^2/4 - mE})f(\vec{q}, \vec{k}; E)}{(p^2 + q^2 + \vec{p}.\vec{q} - mE)(q^2 - k^2 - i\varepsilon)}$$

$$(4.20)$$

The partial wave scattering amplitude, $f_0(p, k; E)$ for s-wave boson-dimer scattering, obtained after using the standard expression

$$f_0(p, k; E) = \frac{1}{2} \int\limits_{-1}^{+1} f(p, k, \cos\theta; E)d(\cos(\theta))$$

and performing the angular integration can be written as

$$f_0(p, k; E) = \frac{8\alpha}{3} K(p, k; E)$$

$$+ \frac{4}{3\pi} \int \frac{q^2 dq(\alpha + \sqrt{3q^2/4 - mE})K(q, k; E)f_0(q, k; E)}{(q^2 - k^2 - i\varepsilon)}, \qquad (4.21)$$

where

$$K(p, k; E) = \frac{1}{pk} \ln\left[\frac{p^2 + k^2 + pk - mE}{p^2 + k^2 - pk - mE}\right] \qquad (4.22)$$

Equation (4.21) along with Eq. (4.22) exhibits a similar structure, apart from the different normalization of the scattering amplitude, as obtained in ref. [40].

Alternatively, we could start from Eq. (2.67), viz.,

$$[\lambda^{-1} - h(3p^2/4 - mE)]F(\vec{p}) = 2 \int d\vec{q} \frac{g(\vec{q} + \vec{p}/2)g(\vec{p} + \vec{q}/2)}{q^2 + p^2 + \vec{q}.\vec{p} - mE - i\varepsilon} F(\vec{q}) \quad (4.23)$$

where

$$h(3p^2/4 - mE) = \int d\vec{q} \frac{g^2(q)}{(q^2 + 3p^2/4 - mE)} \quad (4.24)$$

and applying the boundary condition

$$F(\vec{p}) = (2\pi)^3 \delta(\vec{p} - \vec{k}) + \frac{4\pi a_k(\vec{p}, \vec{k}; E)}{p^2 - k^2 - i\varepsilon} \quad (4.25)$$

we get the equation for the scattering amplitude

$$(3/4)4\pi H\left(\frac{3}{4}p^2 - mE\right) a_k(\vec{p}, \vec{k}; E) = 2(2\pi)^3 \frac{g(\vec{p} + \vec{k}/2)g(\vec{k} + \vec{p}/2)}{(p^2 + k^2 + \vec{p}.\vec{k} - mE)} + 2(4\pi)$$
$$\int d\vec{q} \frac{g(\vec{q} + \vec{p}/2)g(\vec{p} + \vec{q}/2)a_k(\vec{q}, \vec{k}; E)}{(p^2 + q^2 + \vec{p}.\vec{q} - mE)(q^2 - k^2 - i\varepsilon)} \quad (4.26)$$

Again, to get the equation for the scattering amplitude for short-range interaction in EFT, we take the limit $g(p) \rightarrow 1$, and have

$$(3/4)4\pi \left[\frac{2\pi^2}{(\alpha + \sqrt{3p^2/4 - mE})}\right] a(\vec{p}, \vec{k}; E) = 2(2\pi)^3 \frac{1}{(p^2 + k^2 + \vec{p} \cdot \vec{k} - mE)}$$
$$+ 2(4\pi) \int d\vec{q} \frac{a(\vec{q}, \vec{k}; E)}{(p^2 + q^2 + \vec{p} \cdot \vec{q} - mE)(q^2 - k^2 - i\varepsilon)} \quad (4.27)$$

and get the s-wave scattering amplitude

$$\left[\frac{3}{8(\alpha + \sqrt{3p^2/4 - mE})}\right] a_0(p, k; E) = M(p, k; E)$$
$$+ \frac{2}{\pi} \int q^2 dq \frac{M(q, k; E)a_0(q, k; E)}{(q^2 - k^2 - i\varepsilon)}, \quad (4.28)$$

where

$$M(q, k; E) = \frac{1}{2qk} \ln \frac{q^2 + k^2 + qk - mE}{q^2 + k^2 - qk - mE} \quad (4.29)$$

Equation (4.28) is the familiar Skorniakov-Ter Marterosan equation [36] for the three-boson system describing s-wave boson-dimer scattering amplitude.

This similarity in the structure of the integral Eq. (4.21) or (4.28) permits us to employ similar formalism and proceed exactly in the same way as discussed in ref. [40], i.e., by imposing an ultraviolet cutoff Λ to ensure that the integral in the kernel does not diverge and yields results independent of the value of the cutoff used. We shall also employ the same procedure of using the zero-energy scattering length for the n-dimer system as a single subtraction constant to study the behavior of the low-energy scattering amplitude.

We now turn over to set up the formulation for the scattering of n-dimer(n-core(mass,m_c)) scattering at low energies using separable interaction, leading to the coupled one variable integral equations for the scattering amplitude. By allowing the momentum dependent form factors $[g(p)/f(p)]$ to go to one, we shall finally reduce, in the limit, these equations to an effective-field theoretic equations—very much similar to those obtained in ref. [40] for n-d scattering.

Starting from a system of three particles of masses m_1, m_2 of the two neutrons and m_c, the mass of the core with momenta $\vec{p}_1, \vec{p}_2$ and $\vec{p}_3$ respectively, we rewrite the Schrodinger Eq. (2.61) as

$$D(E)\psi = - \sum_{1 \neq j \neq k=1,2,3} m V_{ij}\psi(ij;k),$$

$$D(E) = \vec{p}_{ij}^2/2\mu_{ij} + \vec{p}_k^2/2\mu_{ij-k} - E \quad (i \neq j \neq k = 1, 2, 3) \quad (4.30)$$

and where

$$\vec{p}_{ij} = \frac{m_j\vec{p}_i - m_i\vec{p}_j}{m_i + m_j}; \quad \vec{p}_k = \frac{(m_i + m_j)\vec{p}_k - m_k\vec{p}_j}{m_i + m_j + m_k};$$

$$\mu_{ij} = m_i m_j/(m_i + m_j) \quad \text{and} \quad \mu_{ij-k} = \frac{(m_i + m_j)m_k}{m_i + m_j + m_k}$$

With the s-state interaction for n-n and n-c separable potentials given as

$$V_{12} = -\frac{\lambda_n}{2\mu_{12}}g(p_{12})g(p_{12}'); \quad V_{23} = -\frac{\lambda_c}{2\mu_{23}}f(p_{23})f(p_{23}');$$

$$V_{31} = -\frac{\lambda_c}{2\mu_{31}}f(p_{31})f(p_{31}') \quad (4.31)$$

Following the steps given in Eqs. (2.63)–(2.65), the solution of the Schrodinger Eq. (4.30) for the wave function ψ is written as

$$D(E)\psi = f(\vec{p}_{23})G(\vec{p}_1) + f(\vec{p}_{31})G(\vec{p}_2) + g(\vec{p}_{12})F(\vec{p}_3), \quad (4.32)$$

where the spectator functions, $G(\vec{p})$ and $F(\vec{p})$ satisfy the coupled integral equations:

$$[\Lambda_c^{-1} - h_c(p)]G(\vec{p}) = \int d\vec{q}\, K_1(\vec{p}, \vec{q}; E)G(\vec{q}) + \int d\vec{q}\, K_2(\vec{p}, \vec{q}; E)F(\vec{q})$$

$$[\Lambda_n^{-1} - h_n(\vec{p})]F(\vec{p}) = 2\int d\vec{q}\, K_3(\vec{p}, \vec{q}; E)G(\vec{q}) \qquad (4.33)$$

where the kernels K_1, K_2 and K_3 are explicitly given by

$$K_1(\vec{p}, \vec{q}; E) = \frac{mf(\vec{q} + b\vec{p})f(\vec{p} + b\vec{q})}{[q^2/2a + \vec{q}\cdot\vec{p} + p^2/2a - mE]};$$

$$K_2(\vec{p}, \vec{q}; E) = \frac{mg(\vec{q}/2 + \vec{p})f(\vec{q} + a\vec{p})}{[p^2 + \vec{q}\cdot\vec{p} + q^2/2a - mE]}$$

$$K_3(\vec{p}, \vec{q}; E) = \frac{mg(\vec{q} + \vec{p}/2)f(\vec{p} + a\vec{q})}{[q^2 + \vec{q}\cdot\vec{p} + p^2/2a - mE]} \quad \text{with} \quad a \equiv m_c/(m + m_c),$$

$$b = m/(m + m_c) \text{and} \quad d = (m + m_c)/(2m + m_c) \qquad (4.34)$$

[Note that $m_1 = m_2 = m$ (the neutron mas) and $m_3 = m_c$ (mass of the core nucleus)]. In Eqs. (4.33)

$$\Lambda_c = \lambda_c/2\mu_{23} \quad \text{and} \quad \Lambda_n = \lambda_n/2\mu_{12}$$

$$\text{and} \quad h_c(p) = m\int d\vec{q}\, f^2(q)/[q^2 + p^2/2d - mE];$$

$$h_n(p) = m\int d\vec{q}\, g^2(q)/[q^2 + p^2/2a - mE] \qquad (4.35)$$

To extract out the propagator term from the coefficient of $G(p)$ on the left-hand side in Eq. (4.33), we write

$$[\Lambda_c^{-1} - h_c(p)] = m\left[\int \frac{f^2(q)d\vec{q}}{q^2 + \alpha_{23}^2} - \int \frac{f^2(q)d\vec{q}}{q^2 + ap^2/d - 2maE}\right]$$
$$= m(ap^2/d - 2maE - \alpha_{23}^2)H_c(p),$$

where

$$H_{c(}(p) = \int \frac{f^2(q)d\vec{q}}{(q^2 + \alpha_{23}^2)(q^2 + ap^2/d - 2mE)}$$

The integral $H_c(p)$ can now be simplified in the limit when the form factor f(q) → 1, which reduces to

$$H_c(p)_{f(q)\to 1} \to \frac{2\pi^2}{(\alpha_{23} + \sqrt{ap^2/d - 2mE})}.$$

Using this, the factor $[\Lambda_c^{-1} - h_c(p)]$ is reduced in the limit as

$$[\Lambda_c^{-1} - h_c(p)] = m\frac{a}{d}\left(p^2 - 2md E - \frac{d}{a}\alpha_{23}^2\right)\frac{2\pi^2}{(\alpha_{23} + \sqrt{ap^2/d - 2m E})} \quad (4.36)$$

Since $E = \frac{k^2}{2\mu_1} - \frac{\alpha_{23}^2}{2\mu_{23}} = \frac{k^2}{2md} - \frac{\alpha_{23}^2}{2ma}$, the factor $(p^2 - 2md E - \frac{d}{a}\alpha_{23}^2)$ simplifies to $(p^2 - k^2)$, the inverse of which plays the role of a propagator for an outgoing wave of the neutron relative to the n-core bound state.

The coefficient, $[\Lambda_n^{-1} - h_n(p)]$, of $F(p)$ in Eq. (4.33) can, in a similar fashion, be also simplified as

$$[\Lambda_n^{-1} - h_n(p)] = \left[m\lambda_n^{-1} - m\int d\vec{q}\,\frac{g^2(q)}{q^2 + p^2/2a - m E}\right]$$

In the limit $g(p) \to 1$, the above expression simplifies to

$$[\Lambda_n^{-1} - h_n(p)]_{g(p)\to 1} \to 2\pi^2 m\left(-\frac{1}{a_{nn}} + \sqrt{\frac{p^2}{2a} - m E}\right) \quad (4.37)$$

We now substitute these simplifying factors in the expressions

$$[\Lambda_c^{-1} - h_c(p)] = m\frac{a}{d}\left(p^2 - 2md E - \frac{d}{a}\alpha_{23}^2\right)\frac{2\pi^2}{(\alpha_{23} + \sqrt{ap^2/d - 2ma E})} \quad (4.38)$$

$$[\Lambda_n^{-1} - h_n(p)] = m(2\pi^2)\left(-\frac{1}{a_{nn}} + \sqrt{p^2/2a - m E}\right) \quad (4.39)$$

to write in the spectator functions $G(p)$ and $F(p)$,

$$m\frac{a}{d}(p^2 - k^2)\frac{2\pi^2}{(\alpha_{23} + \sqrt{ap^2/d - 2ma E})}G(\vec{p})$$
$$\equiv \int d\vec{q}\,K_3(\vec{p}, \vec{q}; E)G(\vec{q}) + \int d\vec{q}\,K_2(p, q; E)F(\vec{q}) \quad (4.40)$$

and

$$m(2\pi^2)\left(-\frac{1}{a_{nn}} + \sqrt{p^2/2a - m E}\right)F(\vec{p}) \equiv 2\int d\vec{q}\,K_1(\vec{p}, \vec{q}; E)G(\vec{q}) \quad (4.41)$$

We now apply the boundary conditions

$$G(\vec{p}) = (2\pi)^3\delta(\vec{p} - \vec{k}) + \frac{4\pi f_k(\vec{p})}{(p^2 - k^2 - i\varepsilon)}$$

$$F(\vec{p}) = \frac{4\pi b_k(\vec{p})}{(p^2 - k^2 - i\varepsilon)}$$

to the Eqs. (4.40) and (4.41) and setting the limit $g(p)/f(p) \to 1$, we get the equations for the scattering amplitude for n-dimer scattering

$$\left(\frac{a}{d}\right)\frac{(2\pi^2)}{[\alpha_{23} + \sqrt{ap^2/d - 2maE}]}(4\pi)f_k(\vec{p}) = (2\pi)^3 K_1(\vec{p},\vec{k};E)$$

$$+ (4\pi)\int d\vec{q}\,\frac{K_1(\vec{p},\vec{q};E)f_k(\vec{q})}{(q^2 - k^2 - i\varepsilon)}$$

$$+ (4\pi)\int d\vec{q}\,\frac{K_2(\vec{p},\vec{q};E)b(\vec{q},E)}{(q^2 - k^2 - i\varepsilon)}$$

or $\left(\frac{a}{d}\right)\dfrac{1}{[\alpha_{23} + \sqrt{ap^2/d - 2maE}]}f_k(p) = K_1(\vec{p},\vec{k};E) + \dfrac{1}{2\pi^2}\int d\vec{q}\,\dfrac{K_1(\vec{p},\vec{q};E)f_k(\vec{q})}{(q^2 - k^2 - i\varepsilon)}$

$$+ \frac{1}{(2\pi^2)}\int d\vec{q}\,\frac{K_2(\vec{p},\vec{q};E)b_k(\vec{q})}{(q^2 - k^2 - i\varepsilon)} \qquad (4.42)$$

and

$$\frac{(-1/a_{nn} + \sqrt{p^2/2a - mE})b_k(\vec{p})}{(p^2 - k^2)} = 2K_3(\vec{p},\vec{k};E) + 2(\frac{1}{2\pi^2})\int d\vec{q}\,\frac{K_3(\vec{p},\vec{q};E)f_k(\vec{q})}{(q^2 - k^2 - i\varepsilon)} \qquad (4.43)$$

After performing the angular integration, the expressions (4.42) and (4.43) for the s-wave scattering amplitude get reduced to

$$\left(\frac{a}{d}\right)\frac{1}{[\alpha_{23} + \sqrt{ap^2/d - 2maE}]}f_0(p,k;E) = Z_1(p,k;E)$$

$$+ \frac{1}{\pi}\int q^2 dq\, Z_1(q,k;E)\frac{f_0(q,k;E)}{(q^2 - k^2 - i\varepsilon)};$$

$$+ \frac{1}{\pi}\int q^2 dq\, Z_2(q,k;E)\frac{b_0(q,k;E)}{(q^2 - k^2 - i\varepsilon)} \qquad (4.44)$$

and

$$\frac{(-1/a_{nn} + \sqrt{p^2/2a - mE})b_0(p,k;E)}{(p^2 - k^2)} = 2Z_3(p,k;E)$$

$$+ 2\left(\frac{1}{\pi}\right)\int q^2 dq\,\frac{Z_3(p,q;E)f_0(q,k;E)}{(q^2 - k^2 - i\varepsilon)} \qquad (4.45)$$

where

$$Z_1(p, k; E) = \frac{1}{pk} \ln \frac{p^2/2a + k^2/2a + pk - mE}{P^2/2a + k^2/2a - pk - mE},$$

$$Z_2(p, k; E) = \frac{1}{pk} \ln \frac{p^2 + k^2/2a + pk - mE}{p^2 + k^2/2a - pk - mE},$$

$$Z_3(p, k; E) = \frac{1}{pk} \ln \frac{k^2 + p^2/2a + kp - mE}{k^2 + p^2/2a - kp - mE} \qquad (4.46)$$

Having derived these equations for the n-dimer scattering amplitude within the framework of effective field theory, we defer further discussion on the computational work including the subtraction procedure to the subsequent Chap. 6 on halo nuclei. There we study the bound state properties of ^{20}C as a three-body (n–n–^{18}C) bound state and n–^{19}C scattering using separable potential at low energies and find some exciting results. It would, therefore, be interesting to undertake a comparative study to see, using effective field theory, how far these results are model independent.

Chapter 5
Halo Nuclei: Properties and Experimental Techniques

5.1 Halo Nuclei: A Brief Introduction

What follows in this chapter is a concise report on the elementary aspects of halo nuclei and their experimental studies at Radioactive Ion Beam (RIB) facilities. The primary effort has been to chronicle the salient findings of the experimental endeavor to produce and probe the nuclei near the drip lines. Since the pioneering experiments of Tanihata and collaborators leading to the discovery of the halo structure of ^{11}Li in the mid-eighties, the concerted efforts of both experimentalists and theorists have generated a very large body of knowledge about the halo nuclei. In this chapter, we make no efforts to present a detailed review of this vast field of research. Instead, we would like to provide a brief overview containing the basic motivation for such studies, the salient features of halo nuclei and the rudiments of experimental techniques. The presentation will be mainly centered around light, neutron-rich, drip line nuclei so as to lay down the motivation for their theoretical studies in the following chapter.

Several of the nuclei near the drip lines show exotic halo structures and exhibit exotic properties, hitherto unobserved and unexpected in stable nuclei away from the drip lines. The starting point of any serious discussion on drip line nuclei is the Segre chart of nuclides as shown in Fig. 5.1. The Segre chart is a two-dimensional representation of all possible atomic nuclei in terms of the neutron and proton numbers. This nuclear landscape can also be presented in three dimensions by adding a third axis representing the total ground state energy of each nuclide. The black squares running roughly along the middle of the plot in Fig. 5.1 represent the stable isotopes, around three hundred or so, and form the so-called line of beta stability.

In a three-dimensional plot, they will form the valley (of beta stability) in the potential energy surface. For a given proton (or neutron) number moving along the increasing neutron (or proton) number one reaches the edge of the landscape, the so-called dripline, where the separation energy of the neutron (or proton) becomes zero. The total number of nuclides in the entire region bounded by the two drip

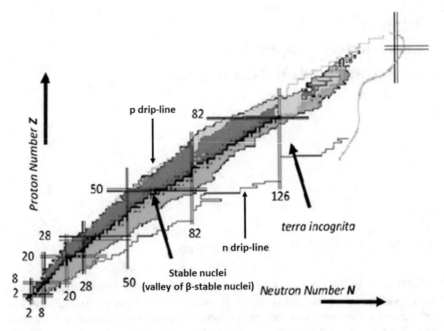

Fig. 5.1 The chart of nuclides. The black dots are known stable isotopes. The colored dots are known unstable nuclei which are either synthesized and studied in the laboratory or are naturally occurring short or long-lived isotopes

lines is expected to be around 7000 or so. However, as of today, the theoretical prediction of the exact number is an outstanding problem. The incomplete nature of our current knowledge about the exact nature of effective nuclear forces inside the nuclear medium, particularly at the drip lines, makes it prohibitively difficult to predict the heaviest nuclide for a given proton or neutron number. Contemporary research in low and medium energy nuclear physics is concerned about the structural properties and reaction dynamics of nuclei in different regions of the Segre chart, away from the line of stability. Globally, the experimental efforts are invested in producing neutron or proton-rich nuclei to reach the drip lines at state-of-the-art RIB facilities.

There are several enduring features of the nuclear structure which have emerged since the beginning of nuclear physics. Namely, the relation that connects the size of a nucleus with the mass number ($R = R_o A^{1/3}$), the nuclear shell structure, the magic numbers for the neutrons or protons providing extra stabilities to the nuclei, etc. As mentioned in the beginning, the nuclei near the drip lines, especially, light neutron-rich nuclei, exhibit unusual structural properties in apparent violation of these long-established features. Very large nuclear matter radius, an apparent halo structure formed by the valence neutron(s), unusually low separation energy of the valence nucleon(s) are some of the unexpected structural features of neutron-rich, light, near-drip-line nuclei. The discovery of the neutron-rich, halo nuclei and their exotic properties are intimately connected with the advent of RIB facilities. The pioneering experiment that formally started the field was performed by Isao Tanihata and collaborators who measured the total interaction cross sections in nucleus-nucleus collisions using secondary beams of stable and unstable Li and Be isotopes [15, 41]. They extracted the interaction radii of these isotopes from the total reaction cross sections. Projectile fragmentation of ^{11}B and ^{20}Ne at 790 MeV/nucleon at the Bevalac facility of Lawrence Berkeley Laboratory produced short-lived nuclei of ^{6,7,8,9,11}Li and ^{7,9,10}Be. The secondary beams so produced, were bombarded on Be, C and Al targets. The total reaction cross sections measured from the depletion of the projectile beam can be connected with the interaction radii of the target and projectile through the equations,

$$I/I_o = e^{-\rho \sigma t} \tag{5.1a}$$

$$\sigma_I = \pi [R_I(P) + R_I(T)]^2 \tag{5.2b}$$

where ρ, σ and t are the target density, total reaction cross-section, and target thickness and $R_I(P)$ and $R_I(T)$ are the interaction radii of the projectile and target, respectively. The startling finding of this path-breaking experiment was the unusually large interaction radius of ^{11}Li, far away from the $R \sim R_o^{1/3}$ behavior (Fig. 5.2). In contrast, the radii of the stable nuclei were as expected and matched with previous electron scattering measurements. At the face of it, this large matter radius of ^{11}Li could have been ascribed to large deformation of the nucleus. In fact, in their paper, Tanihata et al. had concluded that the very large interaction radius of ^{11}Li suggests 'the existence of a large deformation and/or long tail in the matter distribution in ^{11}Li'. A series of seminal experiments followed to measure the charge-changing cross sections [42] and

Fig. 5.2 Plot showing the interaction radii and mass numbers of normal and halo nuclei

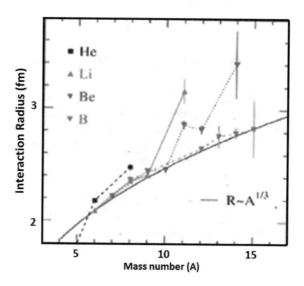

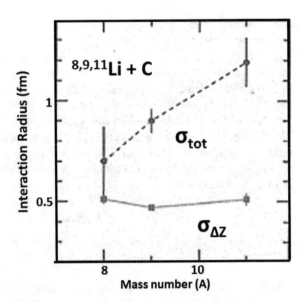

Fig. 5.3 Charge exchange and interaction cross sections for three Li isotopes

quadrupole moments of ^{8,9,11}Li [17]. It was observed that unlike the total interactions the charge-changing cross sections remain same for ^{8,9,11}Li (Fig. 5.3). In addition, experiments at the ISOLDE in CERN [17] showed the quadrupole moments of ^{9}Li and ^{11}Li to be nearly equal. Putting both the results together the obvious conclusion that emerges about the structure of ^{11}Li nucleus is a compact core of ^{9}Li surrounded by a loosely bound veil of two valence neutrons forming a nuclear halo. The term halo was coined in a paper by Hansen and Jonson in 1987 [43]. One can justifiably say the discovery of ^{11}Li and its exotic structure was epochal and ushered in a new era of nuclear physics.

5.2 Salient Structural Features of Halo Nuclei

More than three decades after Tanihata's experiment, the concerted efforts of the nuclear physicists, both theorists and experimentalists, coupled with phenomenal progress in RIB technologies, have resulted in the synthesis of a large number of neutron-rich halo nuclei with structural properties very similar to the ^{11}Li nucleus. We summarize below the striking features of halo nuclei that set them apart from more stable nuclei away from the drip lines. They are,

1. Unusually small separation energy for the last neutron (S_n) or last two neutrons (S_{2n})
2. Very large matter radius
3. Narrow momentum distribution of the fragments of halo nuclei
4. Borromean property of many two-neutron halo nuclei.

Before we discuss these features any further it is in place to summarize the basic conditions for

The halo formation. They are,

1. Small binding energy of the valence nucleon(s)
2. Small orbital angular momentum (most preferably $l = 0$ s-state)
3. Coulomb barrier hinders (proton) halo formation.

The pronounced spatial extent of the halo is obvious from the fact that the binding energies of the halo neutrons are unusually small, enabling them to spread out of the range of the nuclear interaction. It is well known that the separation energy of the last nucleon in a stable nucleus is about 8–10 MeV. For example, the separation energy of the last two neutrons in ^{18}O is about 12.2 MeV. In contrast, the separation energy of the last two neutrons in ^{11}Li is about 370 keV. The valence neutron(s) spread out of the range of the mean-field potential created by the rest of the nucleons forming the core. This is, effectively, some sort of decoupling of the valence nucleon(s) from the compact core. In more standard terms, a halo is said to be formed when more than 50% of the probability density of the nucleon(s) falls outside the range of the potential. In normal beta-stable nuclei, the Fermi energies of the protons and neutrons match. In sharp contrast, there is great asymmetry between the Fermi surfaces in neutron-rich drip line nuclei. The valence neutrons are near the edge of the potential well that makes excitations to the continuum very significant. So, the halo formation is definitely a threshold phenomenon. Most halo nuclei are generally in the ground states of the nuclei and have not more than one bund state. An important exception is that of ^{11}Be which has two bound states. We tabulate a few representative numbers from both 1n- and 2n-halo nuclei in Table 5.1 to bolster what has been said so far about weak bindings of nuclear halos.

Another strikingly important feature of the halo nuclei is the presence of narrow momentum distribution of the fragments. Measurements of momentum distributions of the fragments [both core and neutron(s)] of a halo nucleus on a stable target are integral to the study of loosely bound, halo nuclei. In kinematically complete measurements, both valence nucleons and the core are measured in coincidence. Since very little momentum is transferred in the breakup of the loosely bound projectile moving at high velocity, the momentum distributions of the fragments more or less represent the momentum distributions in the composite projectile nucleus before the breakup. The measured momentum distributions can be either transverse (perpendicular) or

Table 5.1 1-neutron and 2-n separation energies for some halo nuclei. Values are either measured or derived from systematics of Audi and Wapstra [91, 108]

Structure	Nucleus	S_n(keV)	S_{2n}(keV)
1n-halo	^{11}Be	504 (6)	7317 (6)
	^{19}C	160 (120)	4350 (110)
2n-halo	^{6}He	1864 (1)	974 (1)
	^{11}Li	330 (30)	370 (30)
	^{14}Be	1850 (120)	1340 (110)
	^{19}B	1030 (900)	500 (430)

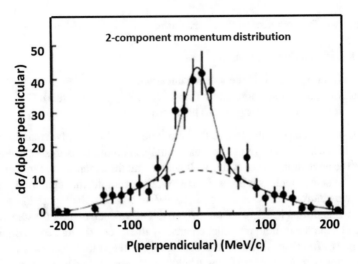

Fig. 5.4 Typical momentum distribution of fragments from breakup of a halo nucleus

longitudinal (parallel) to the beam direction. The narrow momentum distribution of the fragments (the core or the valence neutrons) is a direct manifestation of the Heisenberg uncertainty principle owing to the rather extended spatial structure of the halo. Figure 5.4 represents typical two-component momentum distribution of a halo nucleus from fragmentation experiment. For experimental data of such momentum distribution, we refer to [44].

The condition #2 mentioned above is also easily understandable as large centrifugal barrier would hinder the formation of nuclear halo. The valence nucleon is preferably in an s state with respect to the central core. The significant role of the s wave for neutron halo nucleus like ^{11}Li is discussed and emphasized in the three-body theoretical analysis presented in Chap. 6. In case of proton-halo nuclei, the Coulomb barrier hinders the spatial extension of the halo. Consequently, the proton halos are much less pronounced (or spectacular) than the neutron halos. The well-accepted 1-proton halo nucleus is ^{8}B. Proton removal measurements from ^{8}B have revealed narrow momentum distribution [45]. The width is narrow in comparison with stable nuclei but wider than neutron halo nuclei. The quadrupole moment of ^{8}B is found to be large suggesting a proton halo [46]. ^{17}Ne with 2-proton separation energy of ~0.94 MeV and a relatively narrow 2-proton removal momentum distribution, has been accepted as the lightest 2-proton halo nucleus [47]. The first excited state of ^{17}F is also a possible candidate for proton halo state.

Probably, the most fascinating geometrical property of a typical 2-neutron halo nucleus, like ^{11}Li, is its Borromean nature. It is by now well established that ^{11}Li has a compact core of ^{9}Li and two valence neutrons forming the halo. However, neither the di-neutron nor ^{10}Li is a stable system. In other word, the three-body system of ^{11}Li (core-n-n) is loosely bound but none of the binary systems (n-core or n-n) is bound. This is akin to that of the topological problem of three rings bound together but falling

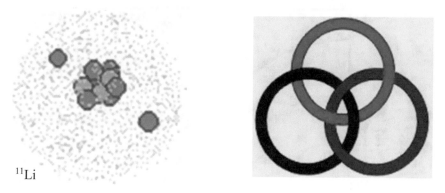

^{11}Li

Fig. 5.5 Artist's impression of the 2-n halo ^{11}Li nucleus (left) and the Borromean rings structure (right) for comparison. The binary sub-systems in both are unbound

apart on removing any one of them (Fig. 5.5). The word Borromean is taken from the heraldic symbol of the Borromeo family in Lago Maggiore island in Italy. It should be said at this point that not all known 2-neutron halo nuclei are Borromean, but some of them. While large fragmentation cross section or very low binding energy of the last neutron(s) hints about the halo structure, it is essential to measure the momentum distributions also to confirm the halo structure of a neutron-rich nucleus.

While the halo structure formed by one or two valence neutrons has drawn maximum interest in this field of research, it is legitimate to ask about what happens in case of multi-nucleon halos. Addition of more neutrons outside the core leads to stronger attraction between the nucleons that prevent formation of a pronounced halo structure as in 1- or 2n- halos. Instead, they give rise to a neutron skin on top of the compact core that does not extend as far as the halo. The often-quoted examples are those of 4, 6, and ^{8}He. While ^{4}He or alpha particle is a highly stable and compact nucleus, the ^{6}He is a 2-neutron halo nucleus (alpha-n-n) with a rather small 2-neutron separation energy (S_{2n}) of about 970 keV. In comparison, the ^{8}He nucleus can be considered as an alpha particle core surrounded by a skin of four valence neutrons. The primary difference between all these three structures lies in their radial matter distributions. It should be noted that the presence of neutron skin or the more extended neutron density distribution than proton is normal in heavy stable nuclei with more neutrons than protons. However, the basic difference between the skin and halo is the much longer range of the halo and its dilute nature. Figure 5.6 presents a typical rendition of matter density distributions for these different cases. For actual experimental data and theoretical calculations, we refer to [48, 49].

The generation of a mean field in a bound nucleus from the two-body interactions of the protons and neutrons is well understood. The emergence of the single-particle shell model states for the mean field and consequently, the magic numbers for protons and neutrons are enduring features of the atomic nucleus. However, a notable feature of the light, drip line nuclei is loss of magic number and emergence of new magic numbers. A rearrangement of the regular shell model states happen in neutron-rich,

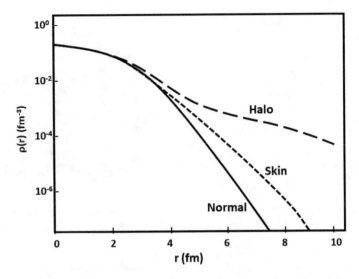

Fig. 5.6 Typical matter density distributions for halo, skin and normal nuclei

halo nuclei and the role of intruder state(s) become very significant. The famous example is that of ^{11}Be, a well established 1-n halo nucleus. A naive shell model picture suggests the ground state spin-parity of ^{11}Be to be $(½)^-$ with the single valence neutron occupying the $0p_{1/2}$ orbital. However, the $1s_{1/2}$ becomes an intruder and drops below the $0p_{1/2}$ state. That explains the $(½)^+$ ground state of ^{11}Be. Another nice example is that of ^{24}O. While simple shell model predicts ^{28}O to be a doubly magic nucleus, it turns out that ^{24}O is indeed a doubly closed magic number nucleus. Both these examples (^{11}Be and ^{24}O) demonstrate the breaking of the $N = 8$ and 20 magic numbers. A fuller discussion on the reason of the change in shell structure is provided in [50]. We would like to end this section on the basic structural properties of halo nuclei by mentioning about another aspect of 2-n halo nucleus, the correlation of the two valence nucleons, which is of great current interest. Experimentally, there are different methods like, determination of matter and charge radii, HBT interferometry, etc. which are used to study the correlation of the di-neutron system [50]. Kinematically complete coincidence measurements of the neutrons and the core of the 2-n halo nucleus provide the information about the n-n and n-core correlations. In the following chapter, we will present a three-body model analysis to reproduce the experimentally measured correlation functions from the breakup of ^{11}Li. Theoretically, it is of much interest to analyze the di-neutron correlation in terms of BCS like (long range) correlation vis-à-vis BEC like short-range correlation. Hagino et al. [51] have studied the di-neutron correlation in ^{6}He, ^{8}He and ^{18}C. They show strong neutron-neutron interaction or significant di-neutron correlation. However, they find the dineutron–dineutron correlation rather weak in ^{8}He and ^{18}C. Strong influence of pairing force on 2-neutron Cooper pair in ^{11}Li has been studied and it has been argued that the influence of pairing is comparatively less in case of the dineutron in ^{6}He [52]. Hagino et al. have also studied correlated 2-neutron emission from unbound ^{26}O [53].

5.3 Known Halo Nuclei and Unbound Nuclei Beyond the Drip Line

As mentioned earlier, the experimental efforts have been to produce radioactive isotopes to expand our known region of the nuclear landscape and reach toward the drip lines. As of now the neutron drip line has been extended up to $Z = 9$ (^{31}F), and $Z = 10$ (^{34}Ne) [54]. Of all the nuclei near the drip lines, synthesized so far, some are halo nuclei. Very recent measurements have confirmed the heaviest bound isotope of boron ^{19}B to be a 2-neutron halo nucleus [55]. Figure 5.7 is a portion of the chart of nuclides presenting the halo nuclei, either confirmed or suggested. Efforts are on to probe the region even beyond the drip lines. The production and studies of unbound nuclei or resonances beyond the drip lines is another fascinating aspect of modern RIB physics. Production of unbound nuclides beyond drip lines is import for better understanding of the structure of halo nuclei. Some of these resonances which have been observed experimentally are, ^{5,7,9,10}He, ^{10}Li, ^{13}Be, ^{16}B. There are also ongoing efforts to observe and study the extremely exotic and super heavy isotopes of hydrogen, namely, 5 and 7H. A very recent finding has been that of ^{11}O resonance which is, interestingly, the mirror of the famous 2-n halo ^{11}Li [56]. Another notable experimental observation has been of the barely unbound isotope of ^{26}O [57]. In recent time, resonance states of most neutron-rich boron isotopes

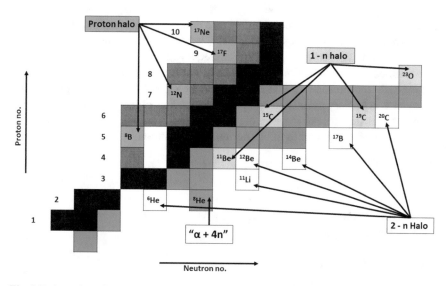

Fig. 5.7 A section of the Segre chart showing the known and suggested halo nuclei

[20,21]B have also been observed from proton removal experiments [58]. For detailed, in-depth discussion on the production and investigation of beyond drip line unbound nuclei, we refer to [50].

5.4 Production of RIB and Major Facilities

The production of any Radioactive Ion Beam is basically a two-step process. A primary beam bombards a target to produce a large number of isotopes. The radioactive isotope to be studied is filtered out and is used as a secondary beam to bombard a second target to carry out specific nuclear reaction. There are two different techniques to produce RIB: Isotope separation on line (ISOL) or target fragmentation and In-Flight separation technique or projectile fragmentation. The fundamental technical challenges are due to the very low production cross section of the RIB, production of a variety of isotopes other than the required one and very short half lives of the isotopes to be studied. It is therefore, imperative to maximize the production cross section by judicious selection of target-projectile combination, very efficient separation of the isotopes of interest and their efficient transportation to the secondary target. All these pose formidable challenges to the accelerator scientists and engineers. In case of ISOL technique, a primary light ion beam of high energy (say proton of several hundred MeV or more) is allowed to bombard a thick and hot target to produce a variety of isotopes by multi-fragmentation of the target. The isotope of interest is diffused out of the hot target and is fed to an ion-source, extracted and further accelerated to produce high energy secondary ion beam. The secondary ion-beam is guided to the secondary target to carry out the nuclear reaction. In case of In-Flight technique, a heavy-ion primary beam at relativistic energy (say ~1 GeV/nucleon) is allowed to get fragmented by bombarding on a thin target. The fragment of interest is separated through a fragment separator and is guided to the secondary target for the nuclear reaction. Remarkable progress in the RIB research over last several decades have resulted in the production of a very large number of short-lived isotopes, both neutron-rich and proton-rich and have helped us expand the known region of the segre chart. For a very recent and comprehensive review of the field, we refer to [59] and references therein. For the sake of completeness, we tabulate below some of the leading ISOL and In-Flight facilities currently in operation.

Major In-Flight Facilities

GANIL	Caen, France
GSI	Darmstadt, Germany
NSCL/MSU	East Lansing, Michigan, USA
RIBF, RIKEN	Tokyo, Japan

Major ISOL Facilities

REX-ISOLDE	CERN, Geneva

TRIUMF, ISAC Vancouver, Canada

Apart from these mega facilities, a large number of relatively smaller but very useful and productive RIB facilities are functional in different parts of the world. Notable among them are, TwinSol setup in Notre Dame University, RIBRAS in Sao Paulo, JYFL in Jyvaskyla, EXCYT in Catania, ALTO in Orsay and CARIBU in ANL. Some of the major international RIB facilities being planned for future or under construction are FAIR at GSI, SPIRAL2 at GANIL, SPES at Legnaro and HIE-ISOLDE at CERN.

5.5 Reaction Studies with Halo Nuclei

The discussion so far has been centered around the structural properties of Halo nuclei. From the beginning of RIB physics the measurements of total reaction cross section, Coulomb dissociation cross sections and momentum distributions in breakup reactions have been the main mode of attack to understand the structure of halo nuclei. Coulomb dissociation of halo nuclei has been particularly active and productive field of research. A variety of techniques have been devised to analyze the data generated from such experiments. Most notable of the findings from Coulomb excitation and dissociation studies has been the pygmy resonance in halo nuclei [60]. Various modes of nuclear reactions, traditionally studied with stable beams, are now also studied with loosely bound halo nuclei. This is a very active field of contemporary research. On nuclear reactions with halo nuclei, we will be necessarily very brief and only mention the types of reactions studied. They are, fusion, transfer, breakup (both Coulomb and nuclear), charge exchange, etc. It is of great contemporary interest to study the angular momentum limit for fusion of halo and weakly bound nuclei. Elastic and inelastic scattering studies (reaction cross sections and angular distributions) of halo nuclei in inverse kinematics on stable targets are also very important and pose challenging problem to standard optical model analysis. The developments in experimental techniques and data reduction have been matched by progress in theoretical reaction studies. For a fuller description of the experimental techniques and theoretical formalisms, we refer to [50, 61, 62, 63].

Chapter 6
Three-Body Approach to Structural Properties of Halo Nuclei and the Efimov Effect

6.1 Introduction

In this chapter, we present a review of the work carried out during the last more than two decades to investigate the structural properties of neutron-rich halo nuclei within the framework of three-body model using separable interactions. The chapter may be broadly divided into two parts. The first part being mainly concerned with the study of structural properties of Borromean-type $2n$-rich halo nuclei like ^{11}Li where, in particular, sufficient experimental data, such as ground state energy, matter radius, momentum distribution and n–n correlation, excited states in ^{11}Li above the breakup threshold and β-decay of ^{11}Li into different channels, are available. The second part is devoted to the search for the occurrence of Efimov states in halo nuclei like ^{14}Be, ^{19}B, ^{22}C and ^{20}C. Specifically, in the case of ^{14}Be nucleus, where we postulated the existence of low-lying virtual state of 2–4 keV in ^{13}Be to predict the binding energy ^{14}Be, the experimental evidence of this low-lying ^{13}Be state was later confirmed by the Michigan group. Among the nuclei studied here, ^{20}C, where the n–^{18}C two-body system has a weak binding to produce the 3-body n–n–^{18}C bound state, is found to be a promising candidate to search for the occurrence of the Efimov states. This warranted us to extend the formulation to study n–^{19}C scattering at low energies to predict Efimov–Fano type resonant states near the three-body threshold. Finally, the renormalization procedure employed in effective field theory equations for the scattering amplitude in n–^{19}C scattering, developed in Chap. 4, is discussed by introducing the three-body scattering length as a subtraction parameter for computing these equations to present a comparative study with those obtained in separable potential model.

6.2 ^{11}Li Halo Nucleus as a Three-Body System: Ground State Properties (Binding Energy, Matter Radius, n–n and n-core Correlations)

One of the earliest examples of halo nuclei which attracted a great deal of attention on the experimental and theoretical fronts is the ^{11}Li nucleus. Experimentally, its abnormally small two-neutron separation energy ($\approx 250\,\mathrm{keV}$), large root mean square radius ($3.14 \pm 0.06\,\mathrm{fm}$) and the existence of a neutron halo are some of the striking features experimentally revealed by measurements of interaction cross sections and by momentum distribution of projectile fragments. On the theoretical side, the conventional shell model and the Hartree–Fock calculations [64, 65] met with only a limited success. On the other hand, there was sufficient experimental evidence to view ^{11}Li nucleus having a cluster-like structure as composed of ^{9}Li core with two neutrons loosely bound to the core. With this picture in view, we [66] proposed, as early as in 1994, a three-body model employing a two-body separable potential to present some preliminary results of the calculations on the basic properties of ^{11}Li nucleus.

The formulation of three-body (n–n–core) system using separable s-state separable potential has already been developed in Sect. 4, where the coupled integral equations for the spectator functions, $G(\vec{p})$ and $F(\vec{p})$, [see Eqs. (4.26)–(4.28)], were derived. This formalism can be easily adapted to the case of ^{11}Li, considered as a three-body system of n, n and structure-less ^{9}Li core. For the n–n potential: $V_{12} = -\frac{\lambda_n}{2\mu_{12}} g(p_{12}) g(p'_{12})$, where $\mu_{12} = m/2 (m_1 = m_2 = m$, mass of the neutron) and λ_n is the strength parameter of n–n interaction and the momentum-dependent function $g(p)$ is taken as $1/(p^2 + \beta^2)$. From the low-energy n–n scattering data, the values of the strength parameter λ_n and the range parameter β are estimated as:

(i) $\lambda_n = 23.43\alpha^3$; $\beta = 6.255\alpha$, which reproduce the value of n–n singlet scattering length, $a_{nn} = -23.69\,\mathrm{fm}$ *and the effective ranger* $r_{nn} = 2.15\,\mathrm{fm}$.
(ii) $\lambda_n = 18.6\alpha^3$, $\beta = 5.8\alpha$ *which gives* $a_{nn} = -23.69\,\mathrm{fm}$ *and* $r_{nn} = 2.32\,\mathrm{fm}$.

As far as the n–^{9}Li interaction, the available experimental data are rather limited: The n–^{9}Li system shows a weakly attractive unbound state of 0.8 MeV. Assuming this to be a virtual state, the parameters $\lambda_c = 14.0\alpha^3$ and $\beta_c = 5.5\alpha$ reproduce the value of n-core scattering length $a_{nc} = -7.7\,\mathrm{fm}$ and $r_{nc} = 2.67\,\mathrm{fm}$.

Thus, starting from the coupled integral Eqs. (4.26) and after performing the angular integrations, we compute these equations numerically, which we solve as an eigenvalue problem. In fact, we feed the value of three-body binding energy E as input and solve the equations as eigenvalue for the two-body strength parameter λ_c. The numerical values so obtained are compared with those determined by the two-body analysis, as depicted in Table 6.1 [66].

It may be noted that the ^{11}Li system even with 200 keV seems to be little over-bound. This may not be surprising if we keep in mind that the singlet n–n interaction used here is oversimplified in the sense that it does not include a repulsive core which should in all probability reduce binding marginally. From the table, it is also clear that

Table 6.1 Parameters of the two-body input potentials

B E of ^{11}Li (MeV)	β/α	λ_n/α^3	β_1/α	λ_c/α^3	λ_c/α^3 from three body
0.34	5.8	18.6	5.0	10.32	12.92
	6.255	23.4	5.0	10.32	12.91
	5.8	18.6	5.5	14.00	17.01
0.20	5.8	18.6	5.0	10.32	12.39
	5.8	18.6	5.5	14.00	16.38

Given the ^{11}Li binding energy, the strength parameter as obtained from the three-body equation is matched with the corresponding value obtained from the two-body analysis

the results appear to be sensitive to the range parameter β_1 for the n ^{9}Li interaction, whereas the corresponding parameter β for the n–n interaction is found to be not so sensitive.

The solutions of the spectator functions $F(P)$ and $G(P)$ satisfying the integral Eq. (4.26) describe the momentum distribution of the 'spectator' particles, i.e., of the ^{9}Li core and of the halo neutron in the presence of the other two particles. Figures 6.1 and 6.2 depict the behavior of $F(p_c)$ and $G(p_1)$ for the core and the halo neutron, respectively, as plotted against their respective momenta in the laboratory frame. The total laboratory momentum $P_L(= p_c + p_1 + p_2)$ was chosen to correspond to a ^{11}Li projectile energy of 66 MeV/nucleon [67]. Experimentally, it has been established that high-energy fragmentation of light nuclei closely reveals the ground state momentum distribution of such fragments. Specifically, a large spatial extent of the halo must be related, via the uncertainty relation, to a narrow momentum spread. It has been experimentally observed that the momentum distribution of ^{9}Li

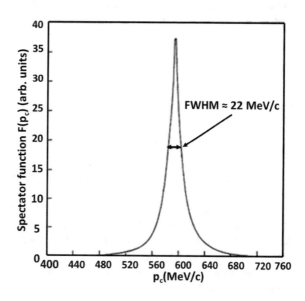

Fig. 6.1 Plot of the spectator function $F(p_c)$ (in arb. Units) versus p_c—the ^{9}Li core momentum in the laboratory frame

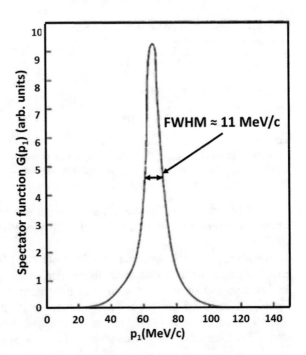

Fig. 6.2 Plot of the spectator function, $G(p_c)$ (in arb. Units) versus p_1—the halo neutron momentum in the laboratory frame

FWHM ≈ 11 MeV/c

expressed in terms of σ varies between 18 and 2 MeV/c. The present results are at least in broad agreement with the conclusions drawn from the experimental findings [68].

From our 'master' three-body Eq. (4.25), we can also make some estimates of the correlation between two halo neutrons or between a halo neutron and ^{9}Li core within the ^{11}Li nucleus. In fact, the term

$$\phi_{nn} = D^{-1}\left(\vec{p}_{12}, \vec{P}_3; E\right)g(p_{12}), \qquad (6.1)$$

which is just the coefficient of $F(P_3)$, can be interpreted as describing the correlation between the two neutrons in momentum space. Similarly, the function

$$\phi_{nc} = D^{-1}\left(\vec{p}_{12}, \vec{P}'_3 E\right)f(\vec{p}_{23}) \qquad (6.2)$$

gives an estimate of the correlation between neutron and ^{9}Li core in the presence of other neutrons.

Experimentally, kinematically complete measurements for Coulomb dissociation of ^{11}Li into ^{9}Li + 2n were made [69] at 28 MeV/nucleon. The data for the correlation function deduced from the relative momentum spectrum between two neutrons in coincidence with ^{9}Li fragment are shown in Fig. 6.3, wherein we also depict a plot of the function ϕ_{nm} versus p_{12}. While the nature of the correlation between two neutrons is yet to be clearly understood, the agreement with the data shows that the

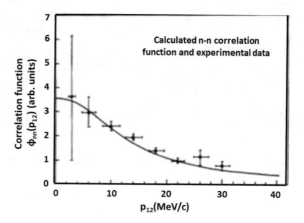

Fig. 6.3 Plot of the correlation function ϕ_{nn} versus P_{12}—the n–n relative momentum (in arb. units.)

present three-body analysis supports the findings of Ieki et al. [69]. By measuring the electromagnetic excitation of ^{11}Li, these authors find that the observed ^{9}Li − $2n$ velocity difference shows a large post-breakup Coulomb acceleration, thereby implying a very short lifetime of the excited state. Thus, they conclude that the soft dipole resonance is not suitable for describing the excitation of ^{11}Li and therefore suggest a direct breakup as a dissociation mechanism.

In Fig. 6.4, a similar plot of n−^{9}Li correlation function ϕ_{nc} versus p_{13} the relative momentum between neutron and core (in arb. units) is shown.

We have also computed, using the functions ϕ_{nn} and ϕ_{nc}, the values of the root mean square radii, r_{nn} and r_{nc}, as a function of three-body binding energy E. These are summarized in Table 6.2.

A comparison of the values for $\bar{r}_{nn}$ and $\bar{r}_{nc}$ obtained here shows that these are somewhat on the higher side. In particular, the analysis of double charge exchange reaction ^{11}B$(\pi^-, \pi^+)^{11}$Li by Gibbs and Hayes [70] puts a strong limit on the size of neutron halo and rules out a value greater than about 6 fm for $\bar{r}_{nn}$. In the present

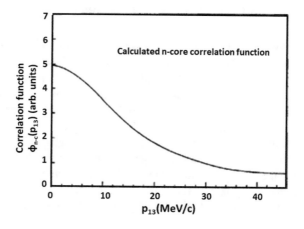

Fig. 6.4 Plot of the correlation function ϕ_{nc} versus P_{13}—the relative momentum between the neutron and the core (in arb. units)

Table 6.2 Values of the root mean square radii of neutron-neutron and neutron–^{9}Li separations calculated using Eqs. (6.1) and (6.2) for different binding energies of ^{11}Li

B.E of ^{11}Li (MeV)	$\bar{r}_{nn}$ (fm)	$\bar{r}_{nn}$ (fm) (from other model calculations [4])	$\bar{r}_{nc}$ (fm)	$\bar{r}_{nc}$ (fm) (from other model calculations [4])
0.20	10.63		10.93	
0.25	9.9		9.86	
0.315	8.93	6.24–7.80	8.87	5.47–6.40

analysis, the inclusion of the repulsive s-state n–n interaction in the spin-singlet state may possibly play a marginal role but it is unlikely to bring about a significant change.

Having calculated $\bar{r}_{nn}$ and $\bar{r}_{nc}$ using three-body model, we can also compute the $\bar{r}_{\text{matter}}$ for ^{11}Li using the relation given by [71], i.e.,

$$\langle r^2 \rangle_{\text{matter}} = \frac{A_c}{A} \langle r^2 \rangle_{\text{core}} + \frac{1}{A} \langle \rho^2 \rangle, \tag{6.3}$$

where $\langle r^2 \rangle_{\text{core}}$ is the mean square value of the radius of ^{9}Li core and the radial variable $\rho^2 = x^2 + y^2$, $\vec{x} = (\vec{r}_1 - \vec{r}_2)/\sqrt{2}$, and $\vec{y} = \sqrt{(2A_c/A)}[(\vec{r}_1 + \vec{r}_2)/2 - \vec{r}_3]$. Working in this representation and taking the values of the ^{9}Li core to be 2.3 fm, we obtain the value of matter radius to be 3.6 fm, which is to be compared with the experimental value 3.14 ± 0.06 fm.

Thus, in this work, we have attempted to extract the structural information about the ^{11}Li nucleus within the framework of three-body model employing separable potentials. We find that even with simplified s-state interaction between different pairs, most of the central gross features of the system can be accounted in reasonable agreement with experimental data.

6.3 Resonant States of ^{11}Li, Probability Distributions and β-decay of Halo Analog States

Having achieved a rather broad qualitative agreement with the experimental data of some of the important properties of ^{11}Li using the above-simplified model, we later carried out detailed studies, in a series of papers [72, 73], extending the model described above to investigate (i) energy positions and widths of resonant states of ^{11}Li above breakup threshold, (ii) probability distributions, momentum distributions and n–n correlations in the ground state, and (iii) β-decay of ^{11}Li to halo analog of ^{11}Be* and to the deuteron $+^9$Li channel. Experiments suggest [74, 75] that β - decay of ^{11}Li into high-lying state of the daughter nucleus ^{11}Be* (18.3 MeV) is the super-allowed Gamow–Teller decay with reduced Gamow–Teller transition probability $B_{\text{GT}} > 1$. Such a large value of B_{GT} implies a large overlap of ^{11}Be* state with the ground state of ^{11}Li. Following Zhukov et al. [76], we considered here ^{11}Be*

state as the isobaric halo analog state consisting of ^{9}Li core with halo neutron–proton pair having isospin $T^h = 0$ and the angular momentum $J^h = 1$.

We first recapitulate some of the essential steps to write down the wave function of ^{11}Li including pair potentials in both **the s- and p-states** of $n-^9$Li core. For the $n–n$ interaction, we choose the $n–n$ singlet $^1S_0$ potential:

$$\langle p'_{12}|V_{12}|p_{12}\rangle = -\frac{\lambda}{2\mu_{12}}g(p_{12})g(p'_{12}), \quad g(p) = \frac{1}{p^2 + \beta^2}, \quad (6.4)$$

where strength parameter $\lambda = 18.6\alpha^3$ and range parameter $\beta = 5.8\alpha$ obtained so as to reproduce the $n–n$ spin-singlet scattering length $a = -23.69$ fm and effective range $r = 2.32$ fm. Similarly, $n-^9$Li separable potential is chosen as

$$\langle p'_{31}|V_{31}|p_{31}\rangle = \sum_{l=0,1} (2l + 1)V^l_{31}(p_{31}, p'_{31})P_l(\cos(\theta))$$
$$V^l_{31}(p_{31}, p'_{31}) = -\frac{\lambda^l}{2\mu_{31}}v^l_{31}(p_{31})v^l_{31}(p''_{31}), \quad v^l_{31}(p) = \frac{p^l}{(p^2+\beta^2_l)^{l+1}} \quad (6.5)$$

with $l = 0, 1$ for the n-core (^{9}Li) interaction operative respectively in s- and p-states to reproduce a virtual $S_{1/2}$ state at an excitation energy of 250 keV [77] and a p-wave resonance observed experimentally at $E_r = 0.538 \pm 0.062$ MeV with resonance width $\Gamma = 0.358 \pm 0.023$ MeV [78]. The range and strength parameters for these potentials are accordingly adjusted to reproduce these observables. These are summarized in Table 6.3.

Using the pair-wise potential given in Eqs. (6.4) and (6.5) in the three-body Schrodinger equation,

Table 6.3 Parameters and observables of pair-wise two-body interactions used in the three-body models of ^{11}Li and ^{11}Be*

Binary system	Partial wave (l)	Strength parameter	Range parameter	Observables
n $-^9$Li	0	$\lambda_0 = 2.25\alpha^3$	$\beta_0 = 3.0\alpha$	$a_s = -13.39$ fm $r_s = 4.93$ fm
n $-^9$Li	1	$\lambda_1 = 2000.0\alpha^5$	$\beta_1 = 5.6\alpha$	$E_r = 0.538$ MeV $\Gamma = 0.475$ MeV
p $-^9$Li	0	$\lambda^0_c = 2.25\alpha^3$	$\beta_0 = 3.0\alpha$	$a_{sc} = -23.98$ fm $r_{sc} = 1.14$ fm
p $-^9$Li	1	$\lambda^1_c = 2000.0\alpha^5$	$\beta_1 = 5.6\alpha$	$E_r = 2.42$ MeV $\Gamma = 6.56$ MeV
n–p	0 (spin-triplet)	$\lambda_{12} = 23.7\alpha^3$	$\beta_{12} = 5.5\alpha$	$a = 5.43$ fm $r = 1.9$ fm

$$D(\vec{p}_{12}, \vec{p}_3; E)\psi(\vec{p}_{12}, \vec{p}_3; E) = -\sum_{ij,k} \int \langle \vec{p}_{ij}|V_{ij}|\vec{p}'_{ij}\rangle \psi(\vec{p}'_{ij}, \vec{p}_k; E) d\vec{p}_{ij}$$

$$(ij, k) = (12, 3), (23, 1), (31, 2) \qquad (6.6)$$

the analytical structure of three-body wave function in the cm system is written as:

$$\psi(\vec{p}_{12}, \vec{p}_3; E) = ND^{-1}(\vec{p}_{12}, \vec{p}_3; E) \begin{bmatrix} g(p_{12})G_1(p_3) + v_{23}^0(p_{23})G_2(p_1) \\ +v_{23}^{(1)}(p_{23})(\hat{p}_{23} \cdot \hat{p}_1)G_3(p_1) + v_{31}^{(0)}(p_{31})G_4(p_2) \\ -v_{31}^{(1)}(p_{31})(\hat{p}_{31} \cdot \hat{p}_2)G_5(p_2) \end{bmatrix}$$

$$(6.7)$$

The spectator functions $G_2(p_1)$ and $G_4(p_2)$ describe the dynamics of one halo neutron when the other halo neutrons and the core are interacting through s-wave. Hence, their structure is exactly the same. Similarly, the functions $G_3(p_1)$ and $G_5(p_2)$, which have the same structure, describe the dynamics of one neutron when the second neutron and the core are interacting through p-wave. The function $G_1(p_3)$ represents the dynamics of ^{9}Li core in the presence of the two neutrons. We thus have three independent functions which satisfy three coupled integral equations. In order to find solutions for these spectator functions, we substitute three-body function Eq. (6.7) in the Schrodinger Eq. (6.6) and compare the similar terms on both sides. This leads to a set of coupled integral equations that can be written in the closed form as

$$G_i(p) = \Lambda_i \left[h_i(p)G_i(p) + \sum_{j=1}^{3} \int d\vec{q}\, K_{ij}(\vec{p}, \vec{q}; E)G_j(\vec{q}) \right], \quad i = 1, 2, 3 \quad (6.8)$$

The detailed structure of constants Λ_i, integrals h_i and the kernels k_{ij} has been given in Ref. [54]. To symmetrize the kernel in Eq. (6.8), we use the following transformations:

$$G_1(p) \to \sqrt{2\Lambda_1}\frac{\chi_1(p)}{p\sqrt{dp}}, G_2(p) \to \sqrt{\Lambda_2}\frac{\chi_2(p)}{p\sqrt{dp}}, G_3(p) \to \sqrt{\Lambda_3}\frac{\chi_3(p)}{p\sqrt{dp}} \quad (6.9)$$

The final symmetric eigen system can now be written as

$$\sum_{j=1}^{3}\sum_{q} K'_{ij}(\vec{p}, \vec{q}; E)\chi_j(q) = \eta(E)\chi_i(p), \quad i = 1, 2, 3 \quad (6.10)$$

where we introduce a parameter $\eta(E)$, which corresponds to the eigenvalue of the kernel of the above integral equation and $K'_{ij}(p, q; E)$ are the integral operators:

$$K'_{ij}(p, q; E) = \Lambda_i h_i(p)\delta_{ij}\delta_{pq} + \sqrt{\Lambda_i \Lambda_j}\sqrt{dp\,dq}\, qp(2\pi)$$

$$\times \int_{-1}^{1} d(\cos\theta) K_{ij}(\vec{p}, \vec{q}; E) \quad i, j = 1, 2, 3 \qquad (6.11)$$

6.3.1   Ground State of ^{11}Li

The condition $\eta(E) = 1$ yields the solution of the integral Eq. (6.8) in the negative energy region and E corresponds to the bound state energy. Equation (6.10) has been numerically solved using Gauss quadrature to compute the three-body ground state energy and the momentum distribution of the spectator functions as eigenvectors. For the ground state of ^{11}Li obtained at energy $E = -0.286$ MeV, the plots of spectator functions versus momentum are shown in Fig. 6.5. We also observe no excited bound state below the three-body breakup threshold in ^{11}Li.

In order to estimate the normalization constant N of the three-body wave function subject to the condition

$$\int \psi(\vec{p}_{12}, \vec{p}_3; E)\psi^*(\overline{p}_{12}, \vec{p}_3; E)d\vec{p}_{12}d\vec{p}_3 = 1 \qquad (6.12)$$

We note that, as a first step, we have to find the analytical structure of these spectator functions which should accurately reproduce the numerical solutions as

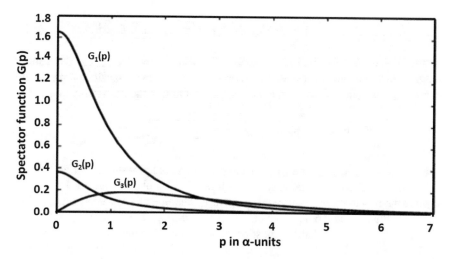

Fig. 6.5 Plot of the spectator functions $G_1(p)$, $G_2(p)$ and $G_3(p)$ for ^{11}Li ground state. Here p along the X-axis is in units of α the deuteron binding energy parameter ($E = \alpha^2/m = 2.226$ MeV)

obtained by solving the integral equation. The analytical expressions which give the best fit are:

$$G_1(p) = \frac{A1}{1 + (p/p_0)^{A2}}, \quad A1 = 1.58\alpha^{-3}, \quad p_0 = 0.9131\alpha, \quad A2 = 2.121 \quad (6.13)$$

$$G_2(p) = G_4(p) = \frac{B1}{1 + (p/p_0)^{B2}}, \quad B1 = 0.343\alpha^{-3}, \quad p_0 = 0.721\alpha, \quad B2 = 2.032 \tag{6.14}$$

$$G_3(p) = G_5(p) = C1(1 - \exp(-p/C2))^{C4} \exp(-(p/C3)),$$
$$C1 = 0.6246\alpha, \quad C2 = 1.3725\alpha, \quad C3 = 1.9530\alpha, \quad C4 = 1.1309 \quad (6.15)$$

With these algebraic representations for the spectator functions as input in three-body wave functions, we compute the quadrature of the integral in Eq. (6.12) to evaluate the normalization constant N, keeping the momentum vectors $\vec{p}_{12}$, and $\vec{p}_3$ as the basis vectors to which all the other vectors such as $\vec{p}_{31}$, $\vec{p}_2$ etc., are properly transformed. The value of N comes out to be $0.0732\alpha^4$. The ground state wave function is thus completely determined up to the normalization constant, which will be used to study the probability distribution, momentum distributions and β-decay to ^{11}Be* state and $d +^9$ Li channel.

6.3.2 Probability Distribution in ^{11}Li

Defining

$$P(\vec{p}_{12}, \vec{p}_{3,}) = \psi(\vec{p}_{12}, \vec{p}_3; E)\psi^*(\vec{p}_{12}, \vec{p}_3; E)d\vec{p}_{12}d\vec{p}_3 \tag{6.16}$$

as the probability distribution of finding ^{9}Li with momentum p_3 in the volume element $p_3^2 dp_3 d\Omega$ and the correlation momentum p_{12} of the two halo neutrons in the element $p_{12}^2 dp_{12}d\Omega$, we can get some insight by plotting the probability density

$$P(p_{12}, p_3) == p_{12}^2 p_3^2 \int d\Omega_{12}d\Omega_3 |\psi(\vec{p}_{12}, \vec{p}_3)|^2 \tag{6.17}$$

versus the momenta p_{12}, and p_{13} in a three-dimensional plot shown in Fig. 6.6.

In Figs. 6.7 and 6.8, we depict the two-dimensional plot of the probability density against the momenta p_{12}, and p_3, respectively, (keeping other momenta fixed). The curves in Figs. 6.7 and 6.8 are the ones that pass through the peak in Fig. 6.6.

These figures show that the most probable values of momenta p_{12}, and p_{13} are 18.28 MeV/c and 30.16 MeV/c, respectively. The full width half maximum (FWHM) Γ_{12} in Fig. 6.7 is 27.42 MeV/c, and by uncertainty principle, this gives the most probable distance between the two halo neutrons as

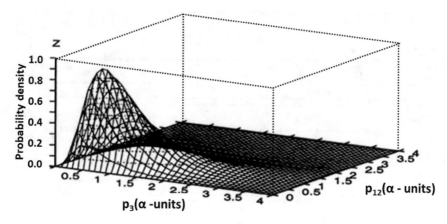

Fig. 6.6 Probability density P along z-axis versus p_{12}, and p_3 (expressed in units of α)

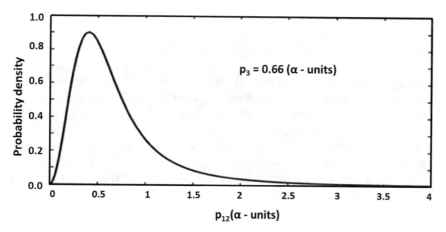

Fig. 6.7 Probability density P (along Y-axis) versus p_3. Here, p_{12}, and p_3 are expressed in units of α

$$r_{12} = r_{nn} = \frac{1}{\Gamma_{12}} = 7.19\,\text{fm} \tag{6.18}$$

Similarly, the FWHM, Γ_3 in Fig. 6.8, is 45.70 MeV/c which gives the most probable value of the distance of ^{9}Li core from the center of mass of the two halo neutrons

$$r_3 = r_{c-nn} = \frac{1}{\Gamma_3} = 4.32\,\text{fm} \tag{6.19}$$

Having calculated r_{12} and r_3 from the three-body model, we can now estimate the rms matter radius, r_{matter}, of ^{11}Li using the relation

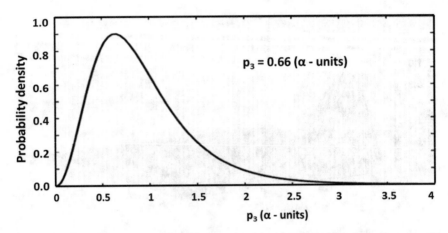

Fig. 6.8 Probability density P (along Y-axis) versus p_3. Here, p_{12}, and p_3 are expressed in units of α

$$\langle r^2 \rangle_{matter} = \frac{A_c}{A} \langle r^2 \rangle_{core} + \frac{1}{A} \langle \rho^2 \rangle, \tag{6.20}$$

originally given by Fedorov et al. [53]. Taking the value of the radius of ^{9}Li and substituting the values of r_{12} and r_3 from Eqs. (6.18) and (6.19) into Eq. (6.20), we obtain the value of the matter radius to be 3.27 fm which is to be compared with 3.2 ± 0.1 fm recently established by Garrido et al. [79].

The study of the behavior of probability amplitude with respect to the angle between vectors $\vec{p}_{12}$ and $\vec{p}_3$ shows that the ground state of ^{11}Li has a maximum probability for a configuration when the n–n relative motion and the recoil motion of ^{9}Li are at right angles to each other in such a way that the halo neutrons when away from the core are closer and they are far away when closer to the core.

6.3.3 n–n Correlations in ^{11}Li from Momentum Distributions

One can also find the behavior of longitudinal momentum distributions of ^{9}Li and that of the halo neutrons by defining $\vec{p} = \vec{p}_{II} + \vec{p}_\perp$. The longitudinal or parallel momentum distribution of ^{9}Li is by definition the momentum distribution of ^{9}Li in the beam direction in fragmentation reaction of ^{11}Li beam on a target at certain beam energy. In the present model, this implies that one should take $\vec{p}_3$ along beam direction only and integrate over the solid angle. Thus, longitudinal momentum distribution of ^{9}Li is given as:

$$\frac{dN}{dp_{3II}} = \int dp_{12} d\Omega_{12} d\Omega_3 p_{12}^2 |\psi(\vec{p}_{12}, \vec{p}_3)|^2 \tag{6.21}$$

This distribution which is compared with the latest experimental data [80] is shown in Fig. 6.9. Here, the overall agreement with the experimental data shows that the inner structure of ^{9}Li core has only marginal role to play at large momenta in so far as the longitudinal component of the momentum distribution is considered. To derive the longitudinal momentum distribution of halo neutron, we first express the three-body wave function in terms of momentum vectors $\vec{p}_{31}$ and $\vec{p}_2$. Then, the longitudinal momentum distribution of halo neutron is simply defined as:

$$\frac{dN}{dp_{II}} = \int dp_{31} d\Omega_{31} d\Omega_2 p_{31}^2 |\psi(\vec{p}_{31}, \vec{p}_2)|^2 \tag{6.22}$$

The longitudinal momentum distribution of halo neutron is shown in Fig. 6.10.

However, the corresponding experimental data, to the best of our knowledge, are not yet available. From Fig. 6.9, the FWHM is estimated to be 46 MeV/c, which is in agreement with the experimental value of 45 ± 3 MeV/c [62]. From Fig. 6.10, FWHM is found to be 35 MeV/c for the halo neutron. Comparison of the two widths shows that

$$\frac{(FWHM)_{9_{Li}}}{(FWHM)_n} = 1.314, \tag{6.23}$$

which is only slightly less than the value for uncorrelated neutron pair, i.e., $\sqrt{2}$ [81]. It can thus be inferred that the two halo neutrons in ^{11}Li are nearly uncorrelated.

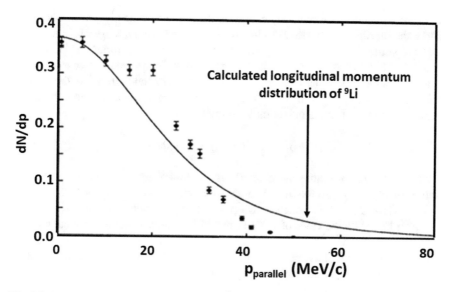

Fig. 6.9 Longitudinal momentum distribution of ^{9}Li (along Y-axis) versus p_{ell}, which is expressed in MeV/c

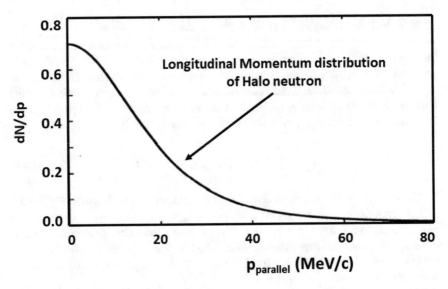

Fig. 6.10 Longitudinal momentum distribution of halo neutron along Y-axis versus $p_{2\parallel}$, which is expressed in MeV/c

6.3.4 Resonant States of ^{11}Li Above the Three-Body Threshold

For positive values of three-body energy E, the kernels of the integral in Eq. (6.10) show singularities on the real axis. In particular, there exists a range of p and q such that the kernel becomes a rapidly varying function of these variables. Hence, the method of bound states cannot be used here. A powerful method known as complex scaling method (CSM) has been recently developed ([82] and references therein) for solving the many body resonances and continuum states. In this method, the linear momenta are transformed into complex numbers as

$$\vec{p} \to \vec{p}\exp(-i\theta), \quad \vec{q} \to \vec{q}\exp(-i\theta) \tag{6.24}$$

The resonance solutions are obtained as a complex energy: $E = E_r - i\Gamma$. The values of E_r and Γ are found to be independent of θ provided $\frac{1}{2}\tan^{-1}(\Gamma/2E_r) < \theta < \theta_c$. Here, E_r is the resonance energy and Γ is the decay width and $\theta_c \approx 0.7$ is an upper bound on the value of θ. Thus, in CSM resonant states are treated in a way similar to the bound states. The first three resonant states of ^{11}Li with the values of the corresponding widths are given in Table 6.4. The available experimental data on the resonance energies are also shown in Table 6.4.

Table 6.4 Resonant states of ^{11}Li above three-body breakup threshold

No.	E_r (MeV), calculated	E_r (MeV), experimental data	Γ (MeV) calculated
1	0.038	0.03 ± 0.04	0.056
2	1.064	1.02 ± 0.07	0.050
3	2.042	2.07 ± 0.12	0.500

6.3.5 Wave Function of Analog ^{11}Be* (18.3 MeV) State in a Three-Body (^{9}Li + n + p) Model

Using the fact that the excess mass of ^{11}Be ground state is 20.174 MeV and the excitation energy of ^{11}Be* is 18.3 MeV, we get separation energy $E_{sp} = 1.8$ MeV relative to the ^{9}Li + n + p threshold in ^{11}Be. The ^{11}Be* (18.3 MeV) state can thus be identified as a bound three-body (^{9}Li + n + p) system with binding energy of 1.8 MeV of the halo neutron and proton pair to the ^{9}Li core. Labeling the neutron, proton and the ^{9}Li core as particles 1, 2 and 3 respectively, the n–^{9}Li interaction is the same as given in the previous section, the additional problem arises in handling the p +^{9} Li system with Coulomb interaction. As is well known, Coulomb potential is not separable. However, Harrington [83] and Cattapan [84] have shown that the problem of determining the Coulomb-corrected phase shifts δ_{cl} due to a short-range separable potential acting between two charged particles is essentially the same as the corresponding problem with neutral particles. The effect of Coulomb potential is completely accounted for by the use of Coulomb-modified potential V^l_{c23} instead of the usual separable potential function. Thus,

$$\langle p_{23}|V_{23}|p_{23}\rangle = \sum_{l=0,1} (2l+1) V^l_{C23}(p_{23}, p'_{23}) P_l(\cos\theta) \tag{6.25}$$

$$V^l_{c23}(p_{23}, p'_{23}) = -\frac{\lambda^l_c}{2\mu_{23}} v^l_{c23}(p_{23}) v^l_{c23}(p'_{23}) \tag{6.26}$$

$$v^l_{c23}(p_{23}) = v^l_{23}(p_{23}) \frac{(2l+1)!}{(2l)!!} C_l(\eta) \exp(2\eta \tan^{-1}(p_{23}/\beta_l)),$$
$$v'_{23}(p) = \frac{p^l}{(p^2+\beta_l^2)^2} \tag{6.27}$$

$$C_l(\eta) = \frac{(l^2+\eta^2)^{1/2}}{(2l+1)} C_{l-1}(\eta), \quad C_0(\eta) = \left|\frac{2\pi\eta}{\exp(2\pi\eta)-1}\right|^{1/2}, \eta(k) = \frac{\mu e_1 e_2}{k} \tag{6.28}$$

In Eq. (6.28), e_1, e_2 denote the two charges, μ is the reduced mass and k is the wave number between the two particles.

In the case of $p -^9$ Li interaction, unfortunately the experimental data particularly at low energies are hardly available. We have therefore to resort to applying the hypothesis of charge independence and assume that the strong interaction part of

$p-^{9}$Li interaction is essentially the same as that of $n-^{9}$Li system. With this proviso, we can then use the parameters of $n-^{9}$Li potential and predict the low-energy observables as influenced by the presence of Coulomb force in $p-^{9}$Li scattering. Employing the potential of Eqs. (6.25)–(6.28), we solve the Schrodinger equation for $p-^{9}$Li system and compute the scattering amplitude in both s- and p-waves. From this, we can then predict the values of the low-energy observables for $p-^{9}$Li scattering as scattering length $a_{sc} = 23.98$ fm and effective range $r_{sc} = 1.143$ fm in the case of s-state interaction while the position and width of the p-wave $p-^{9}$Li resonance are found to be $E_r = 2.42$ MeV and $\Gamma = 6.56$ MeV. Thus, it appears that Coulomb force plays a significant role at such low energies in so far as it affects the values of these observables considerably.

For Gamow–Teller β-*decay* of ^{11}Li involving the halo neutron pair in $T = 1$ and $S = 0$ state, one of the neutrons transforms into a proton resulting in n–p pair in $T = 0$ and $S = 1$ state. For this spin-triplet state of the halo n–p pair, we choose the separable potential

$$\langle p_{12}|V_{12}|p'_{12}\rangle = -\frac{\lambda_{12}}{2\mu_{12}} f(p_{12}) f\left(p'_{12}\right), \quad f(p) = \frac{1}{p^2 + \beta_{12}^2}, \tag{6.29}$$

where the parameters of the potential are given in Table 6.1.

Substituting these potentials for n–p, $n-^{9}$Li *and* $p-^{9}$Li interactions in the three-body Schrodinger Eq. (6.6), the structure of the three-body wave function is written as.

$$\psi_{11_{Be^*}}(\vec{p}_{12}, \vec{p}_3; E) = N_1 D^{-1}(\vec{p}_{12}, \vec{p}_3; E) \begin{bmatrix} f(p_{12})H_1(p_3) + v_{c23}^0(p_{23})H_2(p_1) \\ +v_{c23}^1(p_{23})(\hat{p}_{23} \cdot \hat{p}_1)H_3(p_1) \\ +v_{31}^0(p_{31})H_4(p_2) + v_{31}^1(p_{31})(\hat{p}_{31} \cdot \hat{p}_2)H_5(p_2) \end{bmatrix}, \tag{6.30}$$

where, in Eq. (6.30), $H_i(p)$ ($i = 1-5$) denote the spectator functions which describe the dynamics of the one particle in the presence of the other two mutually interacting particles. The dynamical structure of these functions can be obtained by substituting Eq. (6.30) back into Eq. (6.6) and comparing the corresponding terms on both sides. We thus obtain a set of five coupled integral equations for the functions $H_i(p)$, written in the concise form as:

$$H_i(p) = \Lambda_i \left[h_i(p)H_i(p) + \sum_{j=1}^{5} \int d\vec{q} \, K_{ij}(\vec{p}, \vec{q}; E) H_j(\vec{q}) \right], \quad i = 1-5 \tag{6.31}$$

where

$$\Lambda_1 = \frac{\lambda_{12}}{2\mu_{12}}, \Lambda_2 = \frac{\lambda_0}{2\mu_{31}}, \Lambda_3 = \frac{\lambda_1}{2\mu_{31}}, \Lambda_4 = \frac{\lambda_c^0}{2\mu_{23}}, \Lambda_5 = \frac{\lambda_c^1}{2\mu_{23}}$$

and the elements of the kernel $K_{ij}(\vec{p}, \vec{q}; E)$ are explicitly given in Ref. [73].

Before performing the numerical computation, the kernel of the integral in Eq. (6.31) is symmetrized by using the transformation:

$$H_i(p) \rightarrow \frac{\chi_i(p)}{p\sqrt{dp}}\sqrt{\Lambda_i}, \quad i = 1 - 5. \tag{6.32}$$

The final symmetric eigensystem can now be written by applying the suitable quadrature rule (Gauss–Legendre) to Eq. (6.31) to get

$$\sum_{j=1}^{5}\sum_{q} K'_{ij}(\vec{p}, \vec{q}; E)\chi_j(\vec{q}) = \eta(E)\chi_i(p), \quad i = 1 - 5, \tag{6.33}$$

where we introduce a parameter $\eta(E)$ which represents the eigenvalue of the above equation and $K'_{ij}(p, q; E)$ are the integral operators:

$$K'_{ij}(p, q; E) = \Lambda_i h_i(p)\delta_{ij}\delta_{pq} + \sqrt{\Lambda_i\Lambda_j}\sqrt{dpdq}qp2\pi$$
$$\times \int_{-1}^{+1} d(\cos\theta)K_{ij}(p, q; E), \quad i, j = 1 - 5 \tag{6.34}$$

The numerical solution is achieved by use of the standard library routine 'cg.f' available on the Internet at EISPACK. The solution $\eta(E) = 1$ gives the solution of the three-body energy and the momentum distribution of the spectator functions as eigenvectors. We find that for the input parameters of the two-body potentials the resulting three-body system is slightly overbound. This may possibly be due to the fact that the binary n-p interaction in the spin-triplet case does not include a small admixture of tensor component. It has been found that if we reduce the strength parameter λ_{12} to $22.89\alpha^3$ (less than 3%), it can exactly reproduce the separation energy of the valence n–p pair to 1.8 MeV. For the ^{11}Be* (18.3 MeV) state, the plots of the spectator functions $H_i(p)$ ($i = 1 - 5$) are shown in Figs. 6.11, 6.12, and 6.13.

To estimate the normalization constant of the three-body wave function, we have to determine the analytical structure which should accurately reproduce the numerical solution as obtained from the solution of the integral equation. The algebraic structures that give the best fit are:

$$H_1(p) = \frac{A1}{1 + (p/p_{01})^{A2}}, \quad A1 = 60.920\alpha^{-3}, p_{01} = 0.0760\alpha, A2 = 2.020 \tag{6.35}$$

$$H_2(p) = \frac{B1}{1 + (p/p_{02})^{B2}}, \quad B1 = 0.036\alpha^{-3}, \quad p_{02} = 0.7647\alpha, B2 = 1.918 \tag{6.36}$$

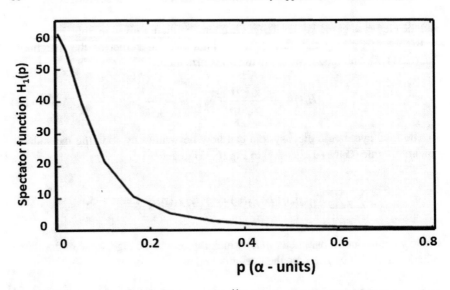

Fig. 6.11 Spectator function $H_1(p)$ versus p for ^{11}Be* excited state at 18.3 MeV. Here, p along X-axis is in units of α

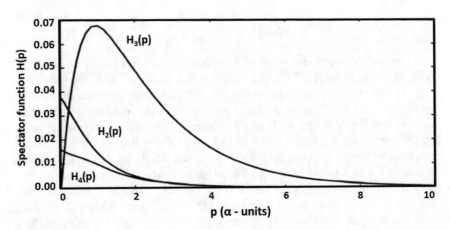

Fig. 6.12 Spectator functions $H_2(p)$, $H_3(p)$, $H_4(p)$ versus p for ^{11}Be* excited state at 18.3 MeV. Here, p along X-axis is in units of α

$$H_3(p) = (C1 \times p) \times \exp(-C2 \times p), \quad C1 = 0.179\alpha^{-3}, C2 = 0.974\alpha^{-1} \quad (6.37)$$

$$H_4(p) = \frac{D1}{1 + (p/p_{03})^{D2}}, \quad D1 = 0.015\alpha^{-3}, p_{03} = 0.7647\alpha, D2 = 2.259$$

$$(6.38)$$

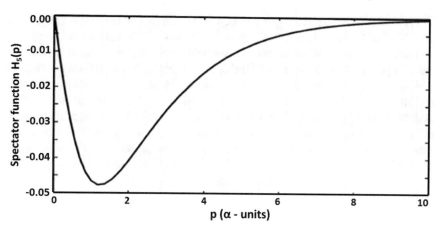

Fig. 6.13 Spectator function $H_5(p)$ versus p for ^{11}Be* excited state at 18.3 MeV. Here, p along X-axis is in units of α

$$H_5(p) = (E1 \times p) \times \exp(-E2 \times p), \quad E1 = -0.101\alpha^{-3}, E2 = 0.800\alpha^{-1}$$
(6.39)

The three-body wave function is then normalized requiring that

$$\int \psi_{11_{Be}}(\vec{p}_{12}, \vec{p}_3; E)\psi_{11Be^*}(\vec{p}_{12}, \vec{p}_3; E)\mathrm{d}\overline{p}_{12}\mathrm{d}\vec{p}_3 = 1$$
(6.40)

Using the algebraic representations for the spectator functions Eqs. (6.35)–(6.39), we compute the quadrature of the integral and determine the value of the normalization constant, $N_1 = 0.157\alpha^4$.

6.3.6   β-Decay of ^{11}Li to Halo Analog ^{11}Be* (18.3 MeV) State

Equipped with the required details to write the complete three-body wave functions for ^{11}Li and ^{11}Be* states, we are now in a position to calculate the transition probabilities of β-decay ^{11}Li to different decay channels. Experimental studies [74, 75, 85, 86] of β-delayed charged particles like deuterons, tritons, ^{4}He and ^{9}Be and neutrons and γ-rays that follow the ^{11}Li β-decay indicate a β-feeding to an excited state at excitation energy of 18.3 MeV in ^{11}Be with a large Gamow–Teller strength, $B_{GT} \geq 1$. Now such a large value of B_{GT} implies a large spatial overlap of the ^{11}Be* state and the ground state of ^{11}Li. Using the three-body wave functions obtained above, we calculate the value of the overlap integral

$$M_{fi} = \langle\psi_{11Be^*}|\psi_{11_{Li}}\rangle = \int \psi_{11Be^*}^*(\vec{p}_{12}, \vec{p}_3; E)\psi_{11_{Li}}(\vec{p}_{12}, \vec{p}_3; E)\mathrm{d}\vec{p}_{12}\mathrm{d}\vec{p}_3$$
(6.41)

Using the wave functions given in Eqs. (6.6) and (6.30), we obtain the value to be 0.51. For Gamow–Teller, β-*decay* of ^{11}Li involving the halo neutron pair in $T = 1$ and $S = 0$ state, when one of the neutrons transforms to a proton resulting in the n–p pair in $T = 0$ and $S = 1$ state, the spin triplet has a statistical weight of 3. Since any of the two halo active neutrons may be undergoing β-*decay*, the total statistical factor should therefore be 6. Thus, the reduced transition probability, B_{GT}, is defined as $B_{GT} = 6|M_{fi}|^2$, which gives the value of 1.5. This appears to be in rather good agreement with the experimental findings. It is important to point out that this is in fact a 'parameter-free' prediction of the present model. As far as the sensitivity of the detailed structure of ^{11}Li *to* β-decay, we find that the admixture of $(p_{1/2})^2$ configuration plays only a marginal role as is evident from the comparison of the spectator functions representing the neutron and ^{9}Li core in s- and p-states (see Fig. 6.5). This is also in qualitative agreement with the conclusions arrived at by Mukha et al. [74] and Borge et al. [75].

6.3.7 β-Decay of ^{11}Li to ^{9}Li + deuteron Channel

The next example to study the sensitivity of the wave function of ^{11}Li is to study β-*decay* of ^{11}Li to the $d-^9$Li channel. The branching ratio for the deuteron channel decay ^{11}Li $\rightarrow^9$ Li $+ d + e^- + \bar{\nu}$ for fixed initial and final states, per unit energy, can be written as [76, 77]:

$$\frac{dB}{dE} = \frac{dW\left(^{11}\text{Li} \rightarrow^9 \text{Li} + d + e^- + \bar{\nu}\right)}{W\left(^{11}\text{Li} \rightarrow^{11} \text{Be}\right)dE}, \tag{6.42}$$

where

$$\frac{dW}{dE} = \left(\frac{G_\beta^2}{8\pi^5}\right)\left(\frac{m_e c^2}{\hbar^4}\right)(\mu k)f(Q - E)B_{GT}(E) \tag{6.43}$$

and

$$W = \left(\frac{G_\beta^2}{2\pi^3}\right)\left(\frac{m_e c^2}{\hbar}\right)\frac{f\left(0^+ \rightarrow 0^+\right)}{t_{1/2}} \tag{6.44}$$

Substituting Eqs. (6.43) and (6.44) in Eq. (6.42), we get the expression for the branching ratio as:

$$\frac{dB}{dE} = \frac{1}{4\pi^2}\frac{\mu k}{\hbar^3}\frac{B_{GT}(E)f(Q - E)t_{1/2}}{ft(0^+ \rightarrow 0^-)} \tag{6.45}$$

$$B_{GT}(E) = 6\lambda^2 \left| \int \psi_d(\vec{p}_{12}) \psi_f(k, \vec{p}_3) \psi_i(\vec{p}_{12}, \vec{p}_3) d\vec{p}_{12} d\vec{p}_3 \right|^2 \qquad (6.46)$$

In Eqs. (6.43) and (6.44), $t_{1/2} = 8.5$ ms is the β-decay half-life of ^{11}Li; k is the wave number of ^{9}Li $+ d$ channel at cm energy E; μ is the reduced mass of the ^{9}Li $+ d$ system; $Q = 2.76$ MeV is the Q-value for the decay of ^{11}Li to the ^{9}Li $+ d$ channel; $\lambda = -1.268$ is the ratio of the axial vector to the vector coupling constant; $f(Q - E)$ is the phase space (Fermi) integral for the deuteron channel and $ft(0^+ \rightarrow 0^-) = 3072.4$ s.

For the deuteron wave function, we use the wave function in momentum space as given in [88]:

$$\psi_d(p_{12}) = \frac{\sqrt{\alpha \beta_1 (\alpha + \beta_1)^3}}{\pi \left(p_{12}^2 + \beta_1^2 \right) \left(p_{12}^2 + \alpha^2 \right)}, \quad \beta_1 = 6.255\alpha, \qquad (6.47)$$

where α is the deuteron binding energy parameter $\left(\alpha^2/m = 2.226 \text{ MeV} \right)$. For the initial state ^{11}Li wave function (ψ_i), we use the normalized wave function (6.7). Here, we consider only the direct decay to the deuteron continuum and assume that due to the low decay Q-value (2.76 MeV), the relative $d-^9$Li motion can be described only by s-wave. The general normalized scattering wave function for $d-^9$Li relative motion can be written as:

$$\psi_f(\vec{p}_3) = \delta\left(\vec{p}_3 - \vec{k}\right) - \frac{1}{2\pi^2} \frac{f_k(p_3)}{k^2 - p_3^2 + i\varepsilon}, \qquad (6.48)$$

where the delta function represents the plane wave part of the $d-^9$Li relative motion and $f_k(p_3)$ denotes the off-shell scattering amplitude to take into account the final state interaction. We need now to construct the scattering amplitude for $d-^9$Li system in the presence of Coulomb interaction. For this, we use the Coulomb-modified function, $V_c(p_3)$, which is constructed similar to the case as discussed for $p-^9$Li interaction and is employed to find the off-shell scattering amplitude appearing in Eq. (6.48).

Thus, using Eq. (6.47) for deuteron wave function, Eq. (6.48) for $d-^9$Li scattering amplitude and Eq. (6.7) for three-body initial state of ^{11}Li, the overlap integral in Eq. (6.46) can be computed to find dB/dE.

The plot of dB/dE versus E is shown in Fig. 6.14, which depicts the maximum probability for deuterons being emitted at 0.32 MeV. Finally, the numerical integration of dB/dE over energy E gives us the total branching ratio to the deuteron channel which is found to be 1.3×10^{-4}. Comparing this with the experimental value of 1.5×10^{-4} [74, 75], we get a rather good agreement with the data.

The main limitations of the present three-body model are: (a) to assume ^{9}Li *core* a structure-less and spinless object and (b) to identify ^{11}Be* (18.3 MeV) state as a component of ^{9}Li $+ n + p$ configuration thereby neglecting some contributions from the complex structure of the ^{11}Be wave function in the excited state. The reasonably

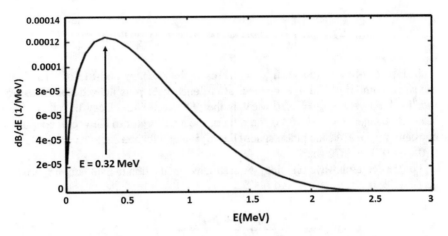

Fig. 6.14 Energy distribution dB/dE versus E. The peak is observed at around 0.32 MeV

good agreement with the available experimental data shows that so long as we restrict ourselves to the low-energy phenomena the above refinements would not significantly change the results of the present analysis. It may be added that an isobaric halo analog state of ^{11}Li *in* ^{11}Be* was investigated by Teranishi et al. [89] using charge exchange reaction ^{11}Li$(p \cdot n)^{11}$Be* in inverse kinematics at $E(^{11}$Li$) = 64$A MeV. The branching ratio for β-decay to deuteron plus ^{9}Li channel was also found to be in good agreement with the experimentally observed value.

It may, however, be emphasized that the present approach has the following distinct advantages: (i) It provides direct information about the momentum distribution of single particles within the nucleus through the knowledge of the spectator functions; (ii) the information about the two-body correlation can be obtained in a rather simple way from the solution of the three-body wave function and (iii) since the three-body observables are computed without resort to any approximation, any discrepancy between the calculated and observed quantity should directly reflect on the limitations of the input two-body potentials used.

6.4   Search for Efimov States in Halo Nuclei like ^{14}Be, ^{19}B, ^{22}C and ^{20}C

While studying the Efimov effect, we found that in a three-body system of resonantly interacting particles when the pair-wise interactions have sufficiently large scattering lengths compared to their respective range, there exists a possibility of an effective three-body potential which may give rise to a sequence of three-body bound states. In the case of two-neutron halos, these nuclei characterized by large spatial extension and very low separation energy of the neutrons may well be regarded as three-body systems which are suited for studying the Efimov effect. One of the promising candidates to investigate the occurrence of Efimov states is the halo ^{14}Be nucleus. With this motive in view, we extended [90] the three-body model developed earlier assuming ^{14}Be as a three-body system comprising of the two halo neutrons and a ^{12}Be–core.

For the neutron–neutron pair, we consider the separable potential:

$$v_{nn} = -\frac{\lambda_n}{2\mu_{nn}} g(p_{nn})g(p'_{nn}), \quad g(p) = \frac{1}{(p^2 + \beta^2)} \tag{6.49}$$

taking the strength parameter $\lambda_n = 18.6\alpha^3$ and the range parameter $\beta = 5.8\alpha$ already considered in Sect. 6.1. In the present analysis, we keep these parameters fixed.

However, for the $n-^{12}$Be binary sub-system, it is worth pointing out that while there is an experimental evidence for a narrow peak in ^{13}Be corresponding to a $d_{5/2}(l = 2)$ resonance unbound by more than 2 MeV [91], a strong possibility for the existence of a low-lying s-orbital state which could have been difficult to identify experimentally cannot be ruled out. It would, therefore, be interesting to explore the consequences of considering such an intruder state in the context of studying ^{14}Be as a three-body system particularly looking for its effect on the possibility of Efimov states. Thus, for the $n-^{12}$Be potential, we set

$$v_{nc} = -\frac{\lambda_c}{2\mu_{nc}} f(p_{nc})f(p'_{nc}), \quad f(p) = \frac{1}{(p^2 + \beta_1^2)} \tag{6.50}$$

and allow the parameters λ_c and β_1 to vary so as to obtain different sets producing virtual and bound two-body systems near zero energy.

The basic structure of the three-body equation in terms of the spectator functions, of $F(p)$ and $G(p)$ the ^{12}Be and of the halo neutrons satisfying the coupled integral equations is essentially the same as given in Eqs. (4.26)–(4.28). However, for the purpose of studying the sensitive computational details of the Efimov effect, we here recast these equations involving only dimensionless quantities. Thus, by defining

$$\tau_n^{-1}(p)F(p) \equiv \phi(p) \quad \text{and} \quad \tau_c^{-1}(p)G(p) \equiv \chi(p), \tag{6.51}$$

where

$$\tau_n^{-1}(p) = \mu_n^{-1} - \left[\beta_r\left(\beta_r + \sqrt{\frac{p^2}{2a} + \varepsilon_3}\right)^2\right]^{-1} \tag{6.52}$$

and

$$\tau_c^{-1}(p) = \mu_c^{-1} - 2a\left[1 + \sqrt{2a\left(\frac{p^2}{4c} + \varepsilon_3\right)}\right]^{-2}, \tag{6.53}$$

where $\mu_n = \pi^2 \lambda_n / \beta_1^3$ and $\mu_c = \pi^2 \lambda_c / (2a\beta_1^3)$ are now dimensionless strength parameters. We also reduce the two coupled integral equations into one equation for $\chi(p)$ by substituting Eq. (4.27) into (4.26) and using the definitions (6.51) for $\phi(p)$ and $\chi(p)$ given above. Thus,

$$\lambda_i \chi(p) = \int d\vec{q}\, K_3(\vec{p}, \vec{q}; \varepsilon) \tau_c(q) \chi(\vec{q})$$
$$+ 2 \iint d\vec{q}\, d\vec{q}'\, K_2(\vec{p}, \vec{q}; \varepsilon) \tau_n(q) K_1(\vec{q}, \vec{q}'; \varepsilon) \tau_c(q') \chi(\vec{q}') \qquad (6.54)$$

where the kernels K_1, K_2 and K_3 are the same as given in Eq. (4.28) except that the variables p, q, etc., are now dimensionless quantities:

$$p/\beta_1 \rightarrow p \text{ and } q/\beta_1 \rightarrow q, \text{ and } -mE/\beta_1^2 \equiv \varepsilon_3,\ \beta_r = \beta/\beta_1. \qquad (6.55)$$

The integral Eq. (6.54) is the eigenvalue equation in λ_i. After having performed the angular integration over the vectors $\vec{q}$ and $\vec{q}'$ and having symmetrized the final kernel, we compute the integral equation as an eigenvalue equation in λ_i.In fact, by feeding the parameters of binary systems, i.e., for n–n and n–c potentials in the right-hand side of Eq. (6.54), we seek the solution of the above equation for the three-body binding energy parameter ε_3 when the eigenvalue λ_i approaches 1, accurate to at least four decimal places. Here, it may be pointed out that the factors τ_n and τ_c defined through Eqs. (6.52) and (6.53) are quite sensitive particularly when the scattering lengths of the binary sub-systems get infinitely large values. In fact, these factors blow up as the variable $p \rightarrow 0$ and the three-body energy parameter ε_3 approaches extremely small values. This necessitates, from the computational point of view, a rather large size of the three-body matrix with double precision so as to minimize the possible truncation errors.

Table 6.5 summarizes the results for the ^{14}Be ground and excited states three-body energy as a result of different two-body input parameters. Keeping the range parameter $\beta_1 = 5.0\alpha$ as fixed,

Table 6.5 ^{14}Be ground and excited states three-body energy for different two-body input parameters

n–^{12}Be Energy keV	λ_1	a_s fm	ε_0 keV	ε_1 keV	ε_2 keV
50	11.71	−21	1350		
5.8	12.32	−61.6	1408	0.053	
2.0	12.46	−105	1450	2.56	0.06
1.0	12.52	−149	1456	3.80	0.22
0.1	12.62	−483	1488	6.10	0.62
0.05	12.63	−658	1490	6.40	0.68
0.01	12.65	−1491	1490	6.90	0.72

We vary the strength parameter λ_1 to produce virtual $n-{}^{12}$Be states at energies varying from 50 to 0.01 keV. The corresponding values of the scattering length range from -21 to -1491 fm. We observe that at 50 keV virtual state, the three-body binding energy for ^{14}Be, as obtained from Eq. (6.54), is 1350 keV which is exactly the experimental value of the two-neutron separation energy. However, this two-body potential does not reproduce any excited state for ^{14}Be. As the virtual state energy of n $-{}^{12}$ Be system is decreased, we not only get the ground state energy, but also the excited state energy for the ^{14}Be system.

In Fig. 6.15a, b, we give a plot of three-body energy versus the two-body scattering length (actually $\ln|a_s|$) for the n $-{}^{12}$ Be system. While the lower curve corresponds to the virtual n $-{}^{12}$ Be states and reproduces the two-neutron separation energy on the lower side, the upper curve corresponds to the higher separation energy directly proportional to the two-body bound state energy. To depict the behavior of the three-body energy, lying in the range from 0 to 2.0 keV versus the two-body scattering length, we use a different scale for the energy and show the plot in Fig. 6.15a.

It is heartening to record that as a follow-up of this analysis experimental study on the search for the evidence of the virtual state of ^{13}Be was undertaken by Thoennessen et al. [74] from Michigan State University. Their investigation did confirm the existence for low-lying s-wave strength of $a_s < -10$ fm representing an unbound state with respect to ^{12}Be and a neutron by <200 keV.

In search of Efimov states, the analysis reported above was further extended [94] to study $2n$ halo nuclei such as ^{19}B, ^{22}C, and ^{20}C where the data on $2n$ separation energies as well as on the n-core bound/unbound states are available [93]. Following exactly the same procedure as outlined above, the integral Eq. (6.54) was numerically computed as an eigenvalue problem. It is worth pointing out that here we are basically encountering a limiting procedure where various factors in the kernels may blow up as the variable $p \to 0$ and the three-body energy parameter approaches extremely small values. As a result, this requires, from the computational point of view, a

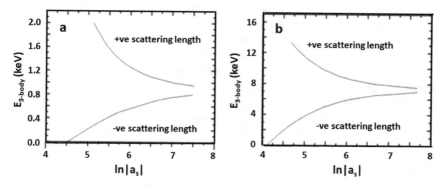

Fig. 6.15 Plot of three-body binding energy vs the two-body scattering length (actually, $\ln a_s|$) for the $n-{}^{12}$Be system **a** shows the behavior of three-body energy up to 2 keV and **b** shows the behavior in high-energy region [see the text]

Table 6.6 Three-body energies for ground and excited states of ^{19}B for different two-body input parameters, $\beta_1 = 4.75\alpha$

$n-^{17}$B energy (keV)	λ_c/α^3	a_s (fm)	ε_0(keV)	ε_1(keV)	ε_2(keV)
514.8	8.49	−6.515	500		
135.3	9.50	−12.71	728		
48	10.0	−21.16	851		
7.7	10.5	−53.2	978	0.16	
0.67	10.75	−179.6	1042	5.40	0.36

Table 6.7 Three-body energy of the ground and excited states of ^{22}C for different two-body input parameters. $\beta_1 = 4.9\alpha$

$n-^{20}$C energy (keV)	λ_c/α^z	a_s(fm)	ε_0(keV)	ε_1(keV)	ε_2(keV)
319	9.82	−8.23	1120		
127	10.5	−13.02	1287		
48.8	11.0	−21.0	1410		
9.3	11.5	−48.2	1540	0.122	
1.46	11.75	−121.5	1608	4.74	0.198

rather large size of the matrix with double precision so as to minimize the possible truncation errors.

Table 6.6 summarizes the results for the nucleus ^{19}B considered as n–n ^{17}B system with ^{17}B as core, whereas in Table 6.7, we present the results for ^{22}C consisting of n–n ^{20}C system with ^{20}C as core.

For the n $-^{17}$ B interaction, we note that $\lambda_c = 8.49\alpha^3$ and $\beta_1 = 4.75\alpha$ give the excitation energy as 514 keV, whereas the experimentally determined value is 530 keV [73]. The excitation energy of 514 keV represents a virtual state with scattering length $a_s = -6.515$ fm and the effective range $r_s = 3.234$ fm. With this two-body interaction along with the n–n potential given earlier, the three-body equation predicts the ground state energy of n–n ^{17}B system as 500 keV which is rather close to the experimental value of 515 keV [91]. However, with these input parameters of the two-body potentials, the three-body equation does not admit any solution for the excited state. When we change the excitation energy for the n–^{17}B system from 514 keV to about 7.7 keV corresponding to a virtual state with scattering length $a_s = -53.2$ fm, only then does the three-body equation start giving the solution for the excited state. In fact, the Efimov region begins to develop when the binary interaction of the n $-^{17}$ B system corresponds to a virtual state with a scattering length of the order of few hundred fermis.

From Table 6.7, it is clear that almost the same scenario holds in the case of ^{22}C considered as n–n ^{20}C system.

For the case of n–^{20}C potential, we choose $\lambda_c = 9.82\alpha^3$ and $\beta_1 = 4.9\alpha$ to fit the excitation energy of 320 keV corresponding to the virtual s state of scattering length $a_s = -8.23$ fm and $r_s = 3.02$ fm. Here also with this potential and the same n–n interaction, the three-body equation predicts the ground state energy of the ^{22}C nucleus as 1120 keV which is in excellent agreement with the experimentally determined value [91]. However, here again the input two-body potential does not allow the three-body equation to admit any solution for the occurrence of the excited states near the breakup threshold. It is only by changing the parameters of the $n - ^{20}$C potential so as to reduce the excitation energy to a few keV or corresponding to scattering length to about a hundred Fermi that the Efimov region begins to develop and the excited states appear for the ^{22}C system.

6.5   Theoretical Analysis/Interpretation for the Efimov States from the Model Equation

Having seen quantitatively the general trend for the appearance of Efimov states in some specific cases, for Borromean-type halo nuclei, it would be instructive to investigate from analytical considerations to see how far such a behavior is universal. For this, it is important to realize that the factors which are crucial in governing the dynamics of binary systems and which play a significant role in describing the kernels of the three-body equation are $\tau_n(p)$ and $\tau_c(p)$ which are defined as

$$\tau_n^{-1}(p) = \mu_n^{-1} - \left[\beta_r \left(\beta_r + \sqrt{\frac{p^2}{2a} + \varepsilon_3} \right)^2 \right]^{-1} \tag{6.56}$$

and

$$\tau_c^{-1}(p) = \mu_c^{-1} - 2a \left[1 + \sqrt{2a \left(\frac{p^2}{4c} + \varepsilon_3 \right)} \right]^{-2}, \tag{6.57}$$

where $\mu_n = \pi^2 \lambda_n / \beta_1^3$, $\mu_c = \pi^2 \lambda_c / (2a\beta_1^3)$, $a = m_c/(m + m_c)$, $c = m_c/(m_c + 2m)$ and $\beta_r = \beta/\beta_1$.

Substituting the above factors, Eq. (6.56) is written more explicitly as

$$\tau_n^{-1} = \frac{\beta_1^3 \lambda_n^{-1}}{\pi^2} - \frac{1}{\beta_r \left(\beta_r + \sqrt{\frac{p^2}{2a} + \varepsilon_3} \right)^2} \tag{6.58}$$

For an unbound system (as, e.g., a virtual state)

$$\frac{\beta^3 \lambda^{-1}}{\pi^2} > 1$$

Since the second term in Eq. (6.58) is a monotonically decreasing function as p increases, this term is always going to be less than the first and the difference between the two would increase as p increases. To express λ_c^{-1} in terms of n–c scattering length, we have the relation

$$\frac{1}{a_{nc}} = \frac{\beta_1}{2} - \frac{\beta_1^4}{2\pi^2\lambda_c} \quad \text{or} \quad \lambda_c^{-1} = \frac{\pi^2}{\beta_1^3}\left(1 - \frac{2}{a_{nc}\beta_1}\right)$$

Thus,

$$\tau_c^{-1} = 2a\left[1 - \frac{2}{a_{nc}\beta_1} - \frac{1}{\left(1 + \sqrt{2a\left(\frac{p^2}{4c} + \varepsilon_3\right)}\right)^2}\right] \qquad (6.59)$$

Clearly, so as long as a_{nc} is negative (representing a virtual state) and the third term is always smaller than the first term, there is hardly any possibility of making $\tau_c^{-1} \to 0$ or $\tau_c \to \infty$, except when a_{nc} approaches a large value and ε_3 goes to the zero limit. This is precisely what has been demonstrated through the numerical analysis. On the other hand, if the binary sub-system is bound corresponding to a positive scattering length, there is a clear possibility of reducing τ_c^{-1} or allowing τ_c to be large enough.

6.6 Occurrence of Efimov States in ^{20}C

A typical example in this case is ^{20}C considered as a bound system of n–n ^{18}C, where $n-^{18}$C is known to be bound with binding energy 160 ± 100 keV [91]. In view of the experimental uncertainties in the determination of the binding energy, we have carried out the calculations for a range of n $- {}^{18}$C binding energies varying from 60 to 200 keV choosing the parameter $\beta_1 = 5.2\alpha$. The results are tabulated in Table 6.8.

From the table, we find that as the binding energy of the $n-^{18}$C increases from 60 to 200 keV, the three-body integral equation predicts the ground state energy

Table 6.8 Ground and excited states three-body energy in ^{20}C for different two-body input parameters. $\beta_1 = 5.2\alpha$

$n-^{18}$C binding energy (keV)	λ_c/α^3	a_s (fm)	ε_0 (keV)	ε_1 (keV)	ε_2 (keV)	N
60	15.51	20.38	3188.03	78.87	65.80	1.01
100	15.89	16.05	3291.54	115.72	100.09	0.94
113.2	16.0	15.15	3317.35	127.41	111.76	0.92
139.60	16.2	13.77	3371.24	150.32	135.29	0.89
168.59	16.4	12.64	3426.03	175.34	163.48	0.86
200	16.6	11.71	3482.95	202.15	194.15	0.84

of ^{20}C increasing from 3.1 to 3.5 MeV in close agreement with the experimental data [91]. In addition, we get excited states as shown in Table 6.8. The first excited states (ε_1) shown in column 5 of the table have binding energies below the n–(nc) scattering threshold ($\varepsilon_1 - \varepsilon_{n-\mathrm{core}}$) which can vary from 2 to 19 keV depending on the $n-^{18}$C input binding energy. In order to analyze these states further, we present the number of bound states (N) for the respective two-body interactions as per Efimov's relations for these quantities [93]. The value of $N \approx 1$ in column 7 is indicative of the occurrence of not more than one Efimov state. We also estimate the size [75] of the first excited state for the different two-body interactions to vary between 50 and 55 fm. The rather large size is also in conformity with expectations for the sizes of Efimov states.

From the three-body equations, we also find an evidence for the occurrence of higher excited states above the n–(nc) scattering threshold but below the three-body breakup threshold. Here, it is interesting to note that the second excited state is below the two-body ($n-^{18}$C) bound state only for 60 keV. At 100 keV, the excited state is just at the breakup threshold. For higher bound state energy, the second excited state is above the two-body breakup. These excited states (ε_2) are also presented in Table 6.8 though this would still need a detailed investigation incorporating the breakup effect explicitly in the structure of the integral equations.

Figure 6.16 presents the plot of the ^{20}C Efimov states (ε_1 and ε_2) relative to the elastic threshold as a function of the $n-^{18}$C binding energy. As discussed earlier, the

Fig. 6.16 Plot of the three-body excited states ε_1 (solid line) and ε_2 (broken line, relative to the elastic threshold versus the $n-^{18}$C binding energy

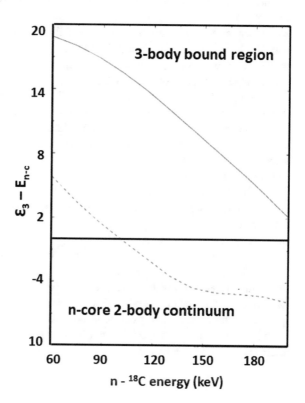

region of 60 keV $< E_{n-c} <$ 200 keV supports the existence of one Efimov state ε_1 (solid line). However, for $E_{n-c} >$ 100 keV the second excited state, ε_2 (broken line), gets destroyed. The conclusions arrived here are in overall qualitative agreement with the analysis of Amorim et al. [95] who also conclude the presence of not more than one Efimov state with an estimated binding energy less than 14 keV below the $n - (nc)$ scattering threshold. The size of the first excited state obtained by us is also in qualitative agreement with the value of Amorim et al. [95] who estimate the size to be at least 35 fm. In conclusion, we find that the $2n$-rich halo nuclei in which the halo neutron is in the intruder low-lying bound state appear to be the most promising candidates to search for the occurrence of Efimov states at energies which may well be within experimental limits to observe.

6.7   Movement of Efimov States in ^{20}C Causing Resonance in $n-^{19}$C Scattering Near Threshold

In 1972, Amado and Noble [96], studying the analytic properties of the Fredholm determinant in a three-boson model, showed that with the increase of potential strength, the Efimov states move into the unphysical sheet associated with the two-body unitarity cut. A detailed analysis was followed by Adhikari et al. [97] to study the movement of the Efimov states in the three-boson model and in the s-wave spin-doublet $(^2S_{1/2})$ three-nucleon system. The key question addressed there was whether the Efimov states, with the increase in potential strength, move over to the virtual states in the unphysical sheet or two of the Efimov states collide to produce a resonance, one of which comes close to the scattering region and produce an observable effect on the physical scattering process.

In the analysis carried in the preceding section, we found the occurrence of Efimov state in ^{20}C near the three-body threshold. For instance, for the $n-^{18}$C binding energy around 140 keV, an Efimov state at about 152 keV was predicted along with the ground state energy of ^{20}C to be 3.18 MeV. We also noticed that if the $n-^{18}$C binding energy is lowered to about 100 keV or even less than that, there could be even more than one bound Efimov states.

In light of further experimental data available on Coulomb dissociation of ^{19}C into $n+^{18}$C studied at 67A [98], where the angular distribution of $n+^{18}$C in the cm system led to the determination of neutron separation energy in ^{19}C to be 530 ± 130 keV, we extended our previous analysis [99] to investigate the behavior of the movement of the Efimov states increasing the $n-^{18}$C binding energy from 200 keV to about 500 keV. Table 6.9 presents the revised estimates of the three-body energy for ^{20}C ground state and excited Efimov states for different two-body (n $-^{18}$ C) binding energies. As noticed earlier, when the n–c pair interaction just binds the two-body (n $-^{18}$ C) system having binding energy around 60–100 keV, the three-body system shows more than one Efimov state but for binding energy equal to or greater than 140 keV, there appears only one Efimov state, the second one moving over to the unphysical

Table 6.9 ^{20}C ground, first excited and second excited state three-body energies for different two-body input parameters

$n-^{18}$C energy ε_2 keV	$\varepsilon_3(0)$ MeV	$\varepsilon_3(1)$ keV	$\varepsilon_3(2)$ keV
60	3.00	79.5	66.95
100	3.10	116.6	101.4
140	3.18	152.0	137.5
180	3.25	186.6	
220	3.32	221.0	
240	3.35	238.1	
250	3.37		
300	3.44		
350	3.51		
400	3.57		
450	3.63		
500	3.69		
550	3.74		

region. When the two-body binding energy increases further, say, beyond 220 keV, even the first Efimov state disappears and moves over to the second unphysical sheet. The difference between the three-body energy of the Efimov state and the two-body binding energy becomes narrower and narrower till it becomes negative with the increase in two-body strength parameter.

To get further insight, we plot in Fig. 6.17 the difference in three-body and two-body binding energy versus the two-body energy for the first and second Efimov states thereby continuously tracing the movement of these states as a function of the two-body binding energy.

We note that while the second Efimov state remains bound till the two-body binding energy ε_2 is 110 keV, it moves over to the continuum (i.e., $\varepsilon_3(2) < \varepsilon_2$) for energy ε_2 greater than that, crossing over the horizontal line representing $\varepsilon_3 = \varepsilon_2$. Similarly, the first Efimov state is bound up to $\varepsilon_3 = 225$ keV and then crosses over the horizontal line at about 230 keV thereby moving over to a continuum state. These 'superfluous' solutions of the three-body homogeneous integral equations provided us a clue to look for the scattering sector to investigate the effect of these states in the continuum.

The feature noted above can, in fact, be attributed to the presence of the singularity in the two-body propagator $\left[\Lambda_c^{-1} - h_c(p)\right]^{-1}$, which can be explicitly written as:

$$\left[\Lambda_c^{-1} - h_c(p)\right]^{-1} = \left[\left(p^2 + 2d\alpha_3^2 - \frac{d}{a}\alpha_2^2\right)h\left(p^2, \alpha_2^2, \alpha_3^2\right)\right]^{-1}, \qquad (6.60)$$

where α_3^2 and α_2^2 are the parameters related respectively to the three-body and two-body binding energies: $E = -\varepsilon_3 = -\alpha_3^2/2\mu_r$, $\varepsilon_2 = -\alpha_2^2/2\mu_{23}$ and $d = (m + m_c)/(2m + m_c)$ and $a = m_c/(m + m_c)$.

Fig. 6.17 Plot of the
difference of the three-body
and two-body binding
energy versus the two-body
binding energy for the first
(solid line) and second
(dashed line) Efimov states

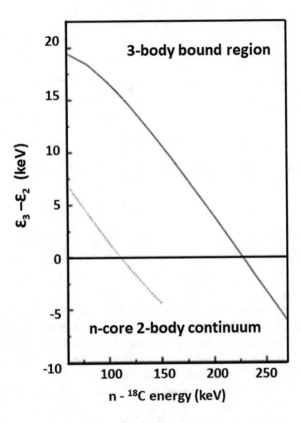

The function $h(p^2, \alpha_2^2, \alpha_3^2)$ represents the form of integral which can be easily worked out. The main point to note here is that the factor $\left(p^2 + 2d\alpha_3^2 - \frac{d}{a}\alpha_2^2\right)^{-1}$ is essentially a two-body propagator. As long as $\alpha_3^2 > \alpha_2^2/2a$, i.e., the three-body energy is numerically greater than that of the two-body, this factor monotonically decreases as p increases. On the other hand, if α_3^2 approaches $\alpha_2^{2a}/2$, we then face a singularity, and for $\alpha_3^2 < \alpha_2^2/2a$, the singularity crosses over to the unphysical sheet. How does such a behavior affect the scattering sector?

Before we move over to the scattering sector, it would be relevant to probe and compare the sensitivity of different sets of the two-body potential parameters on the three-body ground and excited states. Since the experimentally known parameter for ^{19}C halo state is only the two-body binding energy, we have to find these sets which can reproduce the same binding energy of the $n-^{18}$C system. This study may also be quite important to find out how far we can improve upon the Efimov criterion, viz. $a_s \gg r_{\text{eff}}$ by increasing the value of the range parameter β, thereby going to the short-range region, and yet predict the reasonable values of the three-body energies. Thus, choosing a realistic value of $n-^{18}$C binding energy to be 180 keV, we increased the values of β from 5.2α to 7.5α and 10α and obtained the values of strength parameter λ to be $47.31 \alpha^3$ and $109.4 \alpha^3$, respectively. By feeding these values and the

Table 6.10 Comparison of three-body energies (ground, first and second excited states) predicted by different sets of parameters

ε_2, B.E of $n-^{18}$C (keV)	β_1 in units of α	λ_1 in units of α^3	a_s in fm	r_{eff} in fm	ε_3 (G S) (MeV)	ε_{31}	ε_{32}
180	5.2	16.46	12.34	2.273	3.235	172.7 keV	109.6 keV
180	7.5	47.31	12.04	1.615	41.55	410.5 keV	202 keV
180	10.0	109.4	11.67	1.330	155.5	5.700 MeV	618 keV

corresponding values of the parameter β in the expressions for the scattering length and effective range, we found $a_s = 12.04$ fm and $r_{eff} = 1.615$ fm for $\beta = 7.5\alpha$ and $a_s = 11.67$ fm and $r_{eff} = 1.33$ fm for $\beta = 10\alpha$. These values are to be compared with $a_s = 12.34$ fm and $r_{eff} = 2.273$ fm for the set of values, $\beta = 5.2\alpha$ and $\lambda = 16.46\alpha^3$, which we used in the calculations to predict the ground and excited state energies for the three-body ^{20}C system. The first point to note from the two-body parameters is that while the values of range parameter do decrease by increasing the value of β, as expected, so do the values of scattering length parameter. With the result that the ratio, a_s/r_{eff}, only marginally increases. Table 6.10 presents a comparative study of the three-body ground and excited state energies as predicted by the two-body parameters.

It is clear from the table that the three-body ground state energy drastically changes from its realistic value; the corresponding excited state energies also depict a substantial change. The sharp increase in the ground state energy, as we move over the short range, demonstrates that we are probably approaching more toward the region of Thomas collapse rather than to the Efimov region.

Now, we come back to analyze the effect of the singularity pointed above on the behavior of the scattering amplitude for $n-^{19}$C elastic scattering. To study the scattering process, the function $G(p)$ appearing in Eq. (4.26) describing the dynamics of the neutron in the presence of $(n-^{18}$C) system must be subject to the boundary condition:

$$G(\vec{p}) = (2\pi)^3 \delta(\vec{p} - \vec{k}) + \frac{4\pi f_k(\vec{p})}{p^2 - k^2 - i\varepsilon},$$

where the first term represents the plane wave part and the second term is the outgoing spherical wave multiplied by an off-shell scattering amplitude in momentum representation for the scattering of neutron by ^{19}C. The scattering amplitude is normalized such that for s-wave scattering:

$$f_k(\vec{p})\big|_{|\vec{p}|=|\vec{k}|} = \frac{\exp(i\delta)\sin\delta}{k}$$

Before applying the boundary condition, we substitute Eq. (4.28) for $F(p)$ and finally get the equation for the off-shell scattering amplitude as:

$$4\pi\left(\frac{a}{d}\right)h\left(p^2,k^2;\alpha_2^2\right)a_k(p) = (2\pi)^3 K_1(\vec{p},\vec{k}) + 4\pi \int \frac{d\vec{q}\,K_1(\vec{p},\vec{q})a_k(\vec{q})}{q^2 - k^2 - i\varepsilon}$$

$$+ 2(2\pi)^3 \int d\vec{q}\,K_2(\vec{p},\vec{q})K_3(\vec{q},\vec{k})\tau_n(q)$$

$$+ 2(4\pi) \int d\vec{q}\,K_2(\vec{p},\vec{q})\tau_n(q) \int d\vec{q}'\frac{K_3(\vec{q},\vec{q}')a_k(\vec{q}')}{q'^2 - k^2 - i\varepsilon}$$

$$\tag{6.61}$$

This integral equation is to be solved numerically for the scattering amplitude. For s-wave scattering and in the limit when $k \to 0$, the singularity in the two-body cut does not cause any problem; in fact, the amplitude has only the real part. By employing Gauss quadrature and using a mesh size of 80×80 matrix, the off-shell amplitude $a_{k=0}(p)$ is computed by inverting the resultant matrix, which in the limit, $a_0(p)_{p\to0} = -a$, gives the value of $n-^{19}\text{C}$ scattering length. This value, although has a positive sign, turns out to be quite large. Nevertheless, the positive signature of the scattering length not only rules out the possibility of a virtual state in $n-^{19}\text{C}$ system but also supports a bound state, which is quite consistent with the experimental finding.

To investigate the effect of two-body binding energy on three-body scattering length, we study the zero energy $n-^{19}\text{C}$ scattering for different values of the binding energy of the $n-^{18}\text{C}$ system. Thus, by choosing the binding energy $\varepsilon_2 = 100$, 220 and 250 keV, we wish to scan the region where in the first two cases the second Efimov state and the first Efimov state are, respectively, just below the three-body threshold, whereas for the third binding energy both the Efimov states are in the continuum. We find that in all these cases the zero energy scattering length parameter has the value, though quite sensitive to the $n-^{18}\text{C}$ binding energy, yet it retains a positive sign all through. It thus rules out the possibility of the Efimov states turning over to the virtual states in the three-body system. Here, it is important to mention that the investigations [100] carried out so far in studying the behavior of Efimov states in a three-body system, for equal mass particles, both in nuclear and in atomic systems have all shown that the bound Efimov state turns into a virtual state rather than a resonance. As we shall show below, the present study is the first of its kind to show that for unequal mass particles, the Efimov states in the three-body system move over to produce a resonance near the scattering threshold for the scattering of a particle by the bound system.

At incident energies different from zero, i.e., $k \neq 0$, the singularity in the two-body propagator is tackled by following a very elegant technique of continuing the kernel onto a second sheet originally proposed by Baslev and Combes [101]. According to this, the integral contour is deformed from its original position along the real axis to the position rotated by a fixed angle. In other words, we let the variables p, q, etc., become complex by the transformation $p \to p_1 \exp(-i\phi)$ and $q \to q_1 \exp(-i\phi)$. Here, the angle ϕ is merely a parameter to be so chosen that the new contour lies as far away as possible from the singularities. By this operation, the kernel of the integral equation is analytically continued so as to become compact. It must be added that the choice on the values of the angle ϕ is not completely free. These values are restricted by the requirement that the imaginary part of the scattering amplitude calculated from the integral equation must be unitary; i.e., it must satisfy the condition $\text{Im}\left(f_k^{-1}\right) = -k$.

In Fig. 6.18a, b, and c, we depict the behavior of elastic scattering cross section σ_{el} versus incident energy of the neutron on ^{19}C for three different binding energies, i.e., 250, 300 and 350 keV of the $n-^{18}$C system. We find that as the incident energy of the neutron is increased from 1.0 keV (cm energy) to about 1.4 keV, the cross section stays more or less constant at about 100 b. Then, suddenly it shows a sharp rise around 1.6–1.7 keV and falls to about 200 b around 1.8–1.9 keV. The full lines in the figures represent the behavior of the cross section as obtained by computing the integral equation, whereas the dotted curve shows for comparison the fit of the Breit–Wigner resonance shape by using the calculated value of the resonance energy and the width of the resonance. For binding energy 250 keV of the $n-^{18}$C system, the resonance position is obtained at 1.63 keV, while the full width Γ has the value 0.25 keV. However, for higher energies of say, 300 and 350 keV, the position of the resonance shifts to 1.7 and 1.53 keV, respectively, while the resonance width has the corresponding values of 0.27 and 0.32 keV, respectively. Another feature which can be noticed from Fig. 6.18c is that for higher binding energy of the $n-^{18}$C system the behavior of the computed cross section deviates considerably as compared to the Breit–Wigner shape. Nevertheless, there is a definite prediction of the occurrence of a resonance in $n-^{19}$C scattering near threshold at 1.5–1.7 keV.

In order to ensure that the appearance of the peak in the scattering cross section is an unambiguous signature of a resonance in $n-^{19}$C scattering, we apply the following two criteria: (1) Using the relation of the resonance energy $E_{res} = \left(\hbar^2 k_{res}^2\right)/2\mu$, where $k_{res}^2 = k_r^2 - k_i^2$ and k_r and k_i are the real and imaginary parts in the complex k-plane,

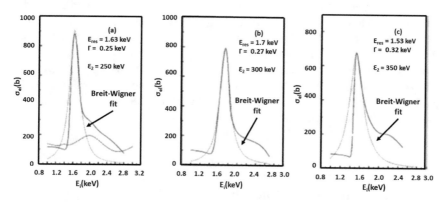

Fig. 6.18 Plot of elastic cross section of $n - ^{19}$C scattering versus cm energy of neutron for $n - ^{18}$C binding energies (ε_2) of **a** 250 keV, **b** 300 keV and **c** 350 keV, respectively. The full curves represent the behavior of the cross sections as obtained by computation, whereas the dotted curves represent the Breit–Wigner fits as described in the text. The dashed curve in (**a**) shows the behavior of the cross section for $n - ^{18}$C binding energy of 200 keV

Fig. 6.19 Argand plot of
$Im(f_k)$ versus $Re(f_k)$ within
a unit circle in the $n - ^{19}C$
scattering when the binding
energy of $n - ^{18}C$ system is
taken as 250 keV

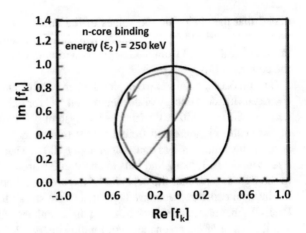

viz. $k = k_r - ik_i$ and the width $\Gamma = (k_r k_i)/\mu$, we compute by using the values of
resonance energy and the width, the corresponding values of k_r and k_i, which are
found to be 3.75×10^{-2} fm^{-1} and 1.53×10^{-3} fm^{-1}, respectively. This ensures that
not only does the resonance lie in the fourth quadrant in the k-plane but also $k_r \gg k_i$.
(2) The second criterion is to study the behavior of $Re(f_k)$ and $Im(f_k)$ on the Argand
diagram as a function of energy around the resonance position. In Fig. 6.19, we depict
a plot of $Im(f_k)$ versus $Re(f_k)$ within a unit circle as we increase the incident energy
of the neutron from 1.0 to 2.75 keV in steps of 0.25 keV, where the binding energy of
the $n - ^{18}C$ system is taken to be 250 keV. We find that as the energy is increased, the
point described by $(Re f_k, Im f_k)$ indicates an anticlockwise motion in the Argand
plot, thereby satisfying an important condition to be obeyed for the occurrence of a
resonance.

It may be interesting to ask whether the scattering cross section would still predict
the resonance-type structure if the pair energy of the bound system is increased to,
say, 200 keV or even less than that in which case the bound Efimov state in the
three-body system is found to occur. To answer this, we depict by a dashed curve the
plot of scattering cross section for $n - ^{19}C$ scattering in Fig. 6.18a taking the $(n - ^{18}C)$
binding energy to be 200 keV. We find that the scattering cross section does not
produce the resonance-like peak structure and remains more or less constant with
the incident energy of the neutron.

6.8 Efimov States and Their Fano Resonances
in a Neutron-Rich Nucleus

In continuation of our investigation studying the appearance of resonances in elastic
scattering cross section as the Efimov states just become unbound, we have attempted
[102] to tie this result to yet another interesting but not fully familiar result that a

general resonance profile is asymmetric, as pointed out by Fano [103] over four decades ago. While such asymmetric profiles have been widely observed in atoms and molecules, resonances in nuclear and particle physics generally show symmetric Lorentzian or Breit–Wigner shapes. It is the combination of the features of the Efimov and Fano phenomena which, in fact, holds promise for the possible observation of such resonances with the asymmetry being used as a diagnostic for the Efimov effect. Our study of the Efimov bound state moving above threshold into the continuum is very much analogous to the recent observation in ultracold cesium which tracked a similar weakening of binding and disappearance as reflected in the loss rate of cesium atoms from an optical trap [104].

In atoms, resonances occur widely due to doubly and multiply excited states, such states often lying in the midst of a continuum built on ground or other low-lying states [105]. Such resonances have also been increasingly common in recent studies of exotic condensed matter systems, where they are sometimes associated with the Kondo phenomenon [106].

Given its asymmetric profile, a resonance is described by three parameters, the energy position E_r, width Γ and a so-called profile index q [104, 105]. Only when this last parameter becomes large does the profile reduce to the symmetric Breit–Wigner form, more familiar to physicists as a resonance, especially in nuclear and particle physics. The reason for this reduction lies in q, which is the ratio of two quantities, the amplitude through the discrete state and the direct amplitude to the underlying continuum. In those instances, when the latter is small or negligible and the embedded discrete state is dominant, the general profile reduces to Breit–Wigner, characterized by just two parameters, E_r and Γ.

The study on $n-{}^{19}C$ elastic scattering predicting a resonance peak in the cross section at energies of a few keV, discussed in the preceding section, showed a clearly asymmetric resonance profile. By tuning parameters for the two-body energy of $n+{}^{18}C$, just above 220 keV when the first excited state of ${}^{20}C$ becomes unbound, the elastic cross section exhibits a resonance as shown in Fig. 6.20. Upon fitting these resonances to the Fano formula [103,105],

$$\sigma = \sigma_0[(q + \varepsilon)^2/(1 + \varepsilon^2)],$$

where $\varepsilon = (E - E_r)/(\Gamma/2)$ is a dimensionless, reduced energy measured from the central position in units of the width, and σ_0 the background cross section far from the resonance, we get the parameters shown in Fig. 6.20.

As pointed out in the earlier section, as we tune the two-body energy beyond 220 keV, the first Efimov state disappears to become quasi-bound state in $n +{}^{19}$ C continuum and we observe the resonance in elastic scattering. A similar effect for the second excited Efimov state was seen to disappear at above 140 keV binding energy and we verified the appearance of a Fano resonance at energy above 140 keV, as shown in Fig. 6.21, pointing to the generality of the phenomenon. Our obtaining the same value of q for the two resonances is in conformity with their constituting a sequence and lends further support to our Efimov interpretation.

Fig. 6.20 Elastic cross
section for n–^{19}C scattering
versus center of mass energy
for a n–^{18}C binding energy
of 250 keV. Calculated cross
section (full curve) is fitted
to the resonance formula in
Eq. (6.1) for three different
values of index parameter q.
The best fit obtained (in our
view) is for $E_r = 1.63$ keV,
$\Gamma = 0.25$ keV parameters.
The dotted line represents a
Breit–Wigner fit to the
calculated curve

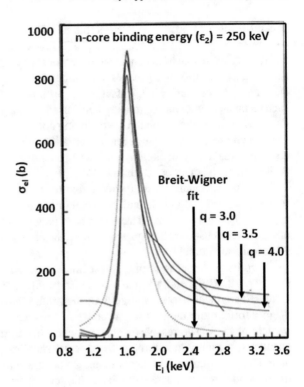

Interestingly, the recent experiment with ultracold atoms [105] also saw only a
couple of Efimov states and again, just as in our study, through their disappearance
as the scattering length was tuned through threshold (in that case, by changing the
magnetic field), without exhibiting the universal exponential scaling with n.

It would be interesting to compare and contrast these nuclear three-body systems
with the somewhat analogous doubly excited states of the helium atom. Figure 6.22
provides such a schematic, singling out the He $2s^2\,^1S$ state as an example [107].

This is a doubly excited state, with both electrons excited out of the ground state
$n = 1$ quantum number. It lies embedded, as shown, in the $1sEs\,^1S$ continuum (E is
the energy of the continuum electron) and mixes with it to give the resonance state.
Note that it lies, approximately, 60 eV above the ground state of helium, which is
enormous in the scale of atomic energies, the first ionization potential of He being
24.6 eV. The resonance would be seen in $(e + He^+)$ scattering at kinetic energies of
about 33 eV. This is in exact analogy with our nuclear example shown alongside in the
figure. Differences between the atomic and nuclear systems are noteworthy. Being an
attractive Coulomb system, He^+ has an excited bound state 2s lying 40.8 eV above
He^+1s, and the $2s^2$ state is bound relative to it (Similarly, there is an infinity of other
doubly excited states below an infinity of excited states of He^+). With short-range
forces between the neutron and ^{18}C, there is no counterpart excited state of ^{19}C and
bound states below it to serve as embedded states in the $n +^{19}$ C continuum.

Fig. 6.21 Similar to Fig. 6.20 with lower binding energies of n–^{18}C, showing a resonance due to the second excited Efimov state for n–^{18}C binding energy of 150 keV. The 150 keV data fit Fano formula with approximately the same Q-value as for the 250 keV curve in Fig. 6.20

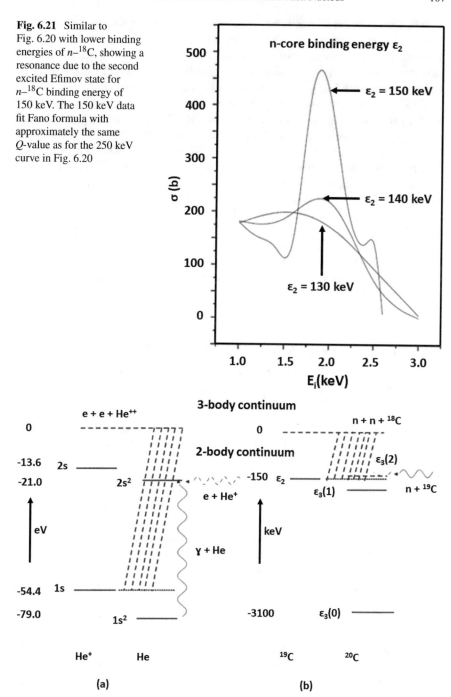

Fig. 6.22 Comparison between He and ^{20}C as three-body systems in atoms and nuclei (For further details, see Fig. 3 in Ref. [102])

It is, therefore, the Efimov effect which provides such a candidate for the subsequent Fano resonance we describe. There is a pleasing interplay between the two phenomena, the loose binding of the Efimov effect being responsible for the substantial overlap between the two pathways that give rise to an asymmetric Fano profile and that asymmetry being the diagnostic for the Efimov phenomenon.

Finally, the analogy provided in the above figure suggests a possible alternative pathway for observing the Fano resonance instead of elastic scattering of neutrons, namely photo-excitation. Just as in the atomic counterpart, where photo-absorption from the ground state with photons of energy 60 eV (approximately) also accesses doubly excited states (although not of $^1 S^e$ *but of* $^1 P^0$ symmetry because of dipole selection rules), we can expect that absorption of gamma rays from lower states such as the ground state of ^{20}C will show Fano resonances from the Efimov states. A further analogy with the observation of doubly excited states through collisions of He with heavy charged particles such as positive ions provides yet another route for observing the Efimov–Fano resonances in nuclei, namely fragmentation of ^{20}C on a heavy target. With the advent of new machines for generating and studying neutron-rich nuclei, we can look forward to the observation in nuclei of the effect which was predicted by Efimov many years ago.

6.9   Generalizing the Study of Efimov Effect in Neutron-Rich (2n-Heavy Core) Nuclei Employing the Above Three-Body Approach

In an attempt to enlarge the scope of present analysis for studying the Efimov effect beyond ^{20}C, we found, on scanning of nuclear data tables, that heavy nuclei such as ^{32}Ne and ^{38}Mg exhibit a structure dynamically similar to ^{20}C [108]. In addition to these halo nuclei, we consider a system of a very heavy core ($A = 100$) with two halo neutrons. Considering a loosely bound binary system of the neutron and the heavy core, we vary the strength of the bound two-body n-core interaction and seek the solutions for the ground and possible excited states.

Extending the above three-body formulation, it should be interesting to find the conditions for the occurrence of the Efimov states in these systems [109] and see how these states evolve into the Feshbach-type resonances. Without elaborating the mathematical steps and the computational details which are similar to what has been discussed in the relevant sections, we present in Table 6.11 the results of the calculations for the three cases, namely mass 102, ^{38}Mg and ^{32}Ne considered as 2n-halo nuclei.

It can be seen from the table that when the n-core pair interaction just binds the two-body system having binding energy between 40 and 60 keV, the three-body system shows more than one Efimov state, the second one being moved over to the unphysical region. When the two-body energy increases further, say beyond 120 keV, even the first Efimov state disappears and moves over to the second unphysical sheet.

Table 6.11 Ground and excited states for the three cases; (i) a nucleus of mass 102 (columns 2, 3, 4), ^{35}Mg (columns 5, 6, 7) and ^{32}Ne (8, 9, 10) for different two-body input parameters

n-core energy ε_2 (keV)	$\varepsilon_3(0)$ keV	$\varepsilon_3(1)$ (keV)	$\varepsilon_3(2)$ (keV)	$\varepsilon_3(0)$ (keV)	$\varepsilon_3(1)$ (keV)	$\varepsilon_3(2)$ (keV)	$\varepsilon_3(0)$ (keV)	$\varepsilon_3(1)$ (keV)	$\varepsilon_3(2)$ (keV)
40	4020	53.6	44.4	3550	61.3	49.9	3420	61.5	50
60	4080	70.4	61.7	3610	80.8	67.1	3480	81.0	67.2
80	4130	86.9	(78.4)	3670	99.2	84.2	3530	99.8	84.3
100	4170	103.1		3710	117	101.4	3570	117.5	101.5
120	4220	(119.3)		3750	134.5	118.9	3620	135	118.9
140	4259			3790	151.6	136.5	3650	152.5	136.7
180	3535			3860	185.6		3730	186.5	
250	4400			3980			3852		
300	4470			4040			3910		
350	4550			4120			3980		

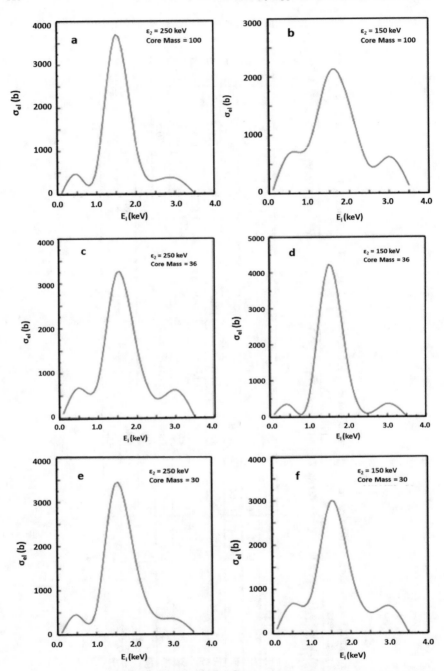

Fig. 6.23 Plot of elastic cross sections for nuclei with a very heavy core of 100 (**a** and **b**), a ^{36}Mg core (**c** and **d**) and a ^{30}Ne core (**e** and **f**)

The overall behavior is very much similar to the results obtained earlier for the ^{20}C nucleus with minor differences in the binding energies of the ground state. In the present case, the ground state appears to be little more strongly bound compared to the $2n$ separation energy in ^{20}C. This can most likely be attributed to the much heavier mass of the core. In addition, the first and second Efimov states move into the continuum at a lower two-body energy compared to that in ^{20}C. In the case of 2-n halo nuclei, like ^{38}Mg and ^{32}Ne, the $2n$ separation energies are comparable to that of ^{20}C (2570 and 1970 keV, respectively) with the n-core systems of ^{37}Mg and ^{31}Ne which are bound by 250 and 330 keV, respectively [90]. The results of the calculations for ^{38}Mg *and* ^{32}Ne are summarized in columns 5, 6, 7 and 8, 9, 10, respectively. For ^{38}Mg, the second excited state disappears in the continuum at around 140 keV n-core energy, while the first excited state disappears at around 220 keV. In the case of ^{32}Ne, same trend continues to appear.

To probe the evolution of the Efimov states into the resonant states, we set out to study these cases in the scattering sector using the formalism developed earlier [99, 102]. Thus, the integral Eq. (6.61) for the scattering amplitude of n-(core $+ n$) scattering below the three-body breakup threshold is computed for these cases. The results of elastic scattering cross sections for scattering by n off the bound n-core systems are depicted in the plots given in Fig. 6.23: for the nucleus with a heavy core of mass 100 [shown in (a) and (b)], by ($n + ^{36}$ Mg as core) shown in (c and d) and by ($n + ^{30}$ Ne as core) depicted in (e and f) for two different n-core interaction energies of 250 keV and 150 keV. A similar asymmetric resonance structure with the centroid around 1.5–1.6 keV and width of around 0.5–0.6 keV is observed in all these cases. This behavior is remarkably similar to that obtained in n $-^{19}$ C elastic scattering. We have also checked that the scattering length of the n-(core $+ n$) system to be positive and large, thereby supporting a bound state. The results presented for the hypothetical nucleus with a heavy core (mass $= 100$) with two valence halo neutrons exhibit a similar behavior to that of realistic 2-n halo nuclei thereby confirming that these features are general enough to persist over a wide mass range and therefore can offer, from the experimental standpoint, a wide range of halo nuclear systems to search for the occurrence of the Efimov states.

6.10 Efimov Effect in Halo Nuclei Using Effective Field Theory

The groundwork for deriving the equations for two- and three-body scattering problem within the framework of effective field theory has already been laid down in the section on effective field theory. In the context of halo nuclear systems, we also set up there the equations for the s-wave scattering amplitude of neutron off the $\left(n - ^{18}\text{C}\right)$ bound state target.

Here, we shall illustrate the subtraction procedure for renormalizing these equations in order to compute the scattering amplitude at finite energies. Let us consider, without any loss of generality, a simpler case of three-boson system describing s-wave boson-dimer scattering amplitude, viz. Eq. (4.28)

$$\left[\frac{3}{8(\alpha + \sqrt{3p^2/4 - mE})} \right] a_0(p, k; E) = M(p, k; k, E)$$

$$+ \frac{2}{\pi} \int_0^\Lambda q^2 dq \frac{M(q, p; k, E)a_0(q, k)}{q^2 - k^2 - i\varepsilon},$$

$$\text{where} \quad M(q, p; k, E) = \frac{1}{2qp} \ln \frac{q^2 + p^2 + qp - mE}{q^2 + p^2 - qp - mE}; \quad mE = \frac{3}{4}k^2 - \alpha^2 \quad (6.62)$$

where we introduce the ultraviolet cutoff parameter, Λ, on the integral in Eq. (6.62). By introducing an ultraviolet cutoff, one finds that while the integral equation can be numerically computed by matrix inversion, its solution is, however, sensitive to the chosen cutoff. One way to resolve this difficulty, as shown by Bedaque et al. [110], is to introduce a one-parameter three-body force term to eliminate the cutoff dependence. There have, however, been alternative approaches to resolve this problem without introducing the three-body force.

Thus, for instance, Blankenleider and Gagelia [111] suggest to construct the actual physical solution of Eq. (6.62), consistent with the unitarity of the scattering amplitude, by performing a straightforward numerical analysis with different cutoff parameters. The second possibility to remove these unphysical solutions from the kernel of Eq. (6.62) is through one subtraction, resulting in an equation which is renormalization group invariant.

To see this, let us write Eq. (6.62) for the scattering threshold by setting $k = 0$, to get

$$\left[\frac{3}{8(\alpha + \sqrt{3p^2/4 + \alpha^2})} \right] a_0(p) = \frac{1}{p^2 + \alpha^2} + \frac{2}{\pi} \int_0^\Lambda dq M(q, p; 0, \Lambda)a_0(q),$$

$$(6.63)$$

Note that $a_0(p)$ is still an off-shell quantity for $p \neq 0$; only at $p = 0$, it becomes on-shell and is related to three-body scattering length, a_3; viz. $a_0(p)|_{p=0} = -1/a_3$. Writing Eq. (6.63) for $p = 0$ and subtracting the resulting equation from (6.63), we get

$$\left[\frac{3}{8\left(\alpha + \sqrt{3p^2/4 + \alpha^2}\right)} \right] a_0(p) - \frac{3}{16\alpha} a_0(0) = \frac{1}{p^2 + \alpha^2} - \frac{1}{\alpha^2} + \frac{2}{\pi}$$

$$\int_0^\Lambda dq \left[\frac{1}{2pq} \ln\left(\frac{q^2 + p^2 + qp + \alpha^2}{q^2 + p^2 - qp + \alpha^2} \right) - \frac{1}{q^2 + \alpha^2} \right] a_0(q),$$

$$(6.64)$$

It has been explicitly demonstrated by Hammer and Mehen [112] by operating $\Lambda d/d\Lambda$ on both sides of Eq. (6.64) that the subtracted equation has a unique solution which is renormalization group invariant. This analysis can be extended to the integral equation for half-off-shell amplitude at any energy. Basically, this is the line of approach employed in [40] which we are extending here in the case of n $-^{19}$ C scattering.

Starting from Eqs.(4.37) and (4.38), we redefine, for the sake of convenience, the off-shell scattering amplitudes, $f_0(p, q; E)$ and $b_0(p, q; E)$ as:

$$(a/d)\frac{1}{\left[\alpha_c + \sqrt{ap^2/d - 2maE}\right]} f_0(p, q; E) = \tau(p; E) f_0(p, q; E) \equiv T(p, q; E)$$

$$(6.65)$$

and

$$\left[-1/a_{nn} + \sqrt{p^2/2a - mE}\right] b_0(p, q; E) = \chi(p; E) b_0(p, q; E) \equiv B(p, q; E),$$

$$(6.66)$$

where $\tau(p; E) = (a/d)\frac{1}{\left[\alpha_c+\sqrt{ap^2/d-2maE}\right]}$ and $\chi(p; E) = \left[-1/a_m + \sqrt{p^2/2a - mE}\right]$ and α_c is the binding energy parameter for n-core BE, $\varepsilon_c = -\alpha_c^2/2ma$ and a_{nn} is the neutron – neutron scattering length.

Substituting Eq. (4.38) into Eq. (4.37) and using Eqs. (6.65) and (6.66), we write the off-energy shell amplitude:

$$T(p, q; E) = Z_1(p, q; E) + \frac{1}{\pi} \int q'^2 dq' Z_1(p, q'; E) \frac{\tau^{-1}(q'; E) T(q', q; E)}{(q'^2 - k^2 - i\varepsilon)}$$

$$+ \frac{2}{\pi} \int q'^2 dq' Z_2(p, q'; E) \chi^{-1}(q'; E) Z_3(q', q; E)$$

$$+ \frac{2}{\pi^2} \int q'^2 dq' Z_2(p, q'; E) \chi^{-1}(q'; E)$$

$$\int q''^2 dq'' \frac{Z_3(q', q'', E) \tau^{-1}(q''; E) T(q'', q; E)}{(q''^2 - k^2 - i\varepsilon)}$$

$$(6.67)$$

where the expressions Z_1, Z_2 and Z_3 have been defined in (4.39).

Consider Eq. (6.67) for half-shell scattering amplitude when $q = k = 0$ so that in that case $E = \varepsilon_c$ and Eq. (6.67) reduces to:

$$T(p, 0; \varepsilon_c) = Z_1(p, 0; \varepsilon_c) + \frac{1}{\pi} \int dq' Z_1(p, q'; \varepsilon_c) \tau^{-1}(q'; \varepsilon_c) T(q', 0; \varepsilon_c)$$

$$+ \frac{2}{\pi} \int q'^2 dq' Z_2(p, q'; \varepsilon_c) \chi^{-1}(q'; \varepsilon_c) Z_3(q', 0; \varepsilon_c)$$

$$+ \frac{2}{\pi^2} \int q'^2 dq' Z_2(p, q'; \varepsilon_c) \chi^{-1}(q'; \varepsilon_c)$$

$$\int dq'' Z_3(q', q''; \varepsilon_c) \tau^{-1}(q''; \varepsilon_c) T(q'', 0; \varepsilon_c) \tag{6.68}$$

The kernel in the integral Eq. (6.68) is known to diverge at large cutoff values and does not yield a unique solution. However, if we consider the on-shell amplitude at the threshold, i.e., at $p = 0$, $q = k = 0$ and $E = \varepsilon_c$, Eq. (6.68) takes the form:

$$T(0, 0; \varepsilon_c) = Z_1(0, 0; \varepsilon_c) + \frac{1}{\pi} \int dq' Z_1(0, q'; \varepsilon_c) \tau^{-1}(q'; \varepsilon_c) T(q', 0; \varepsilon_c)$$

$$+ \frac{2}{\pi} \int q'^2 dq' Z_2(0, q'; \varepsilon_c) \chi^{-1}(q'; \varepsilon_c) Z_3(q', 0; \varepsilon_c)$$

$$+ \frac{2}{\pi^2} \int q'^2 dq' Z_2(0, q'; \varepsilon_c) \chi^{-1}(q'; \varepsilon_c)$$

$$\int dq'' Z_3(q', q''; \varepsilon_c) \tau^{-1}(q''; \varepsilon_c) T(q'', 0; \varepsilon_c)$$

$$= -\tau(0; \varepsilon_c) A_s, \tag{6.69}$$

where A_s is the scattering length of the scattering of neutron by ^{19}C composed of bound $(n + ^{18}C)$ system. Since the value of A_s for $n{-}^{19}C$ scattering length is not known experimentally, we here use the value we determined earlier employing the separable potential model.

Subtracting Eq. (6.69) from (6.68), we then get

$$T(p, 0; \varepsilon_c) + \tau(0, \varepsilon_c) A_s = \Delta Z_1(p, 0; 0, 0; \varepsilon_c)$$

$$+ \frac{1}{\pi} \int dq' \Delta Z_1(p, q'; 0, q'; \varepsilon_c) \tau^{-1}(q'; \varepsilon_c) T(q', 0; \varepsilon_c)$$

$$+ \frac{2}{\pi} \int q'^2 dq' \Delta Z_2(p, q'; 0, q'; \varepsilon_c) \chi^{-1}(q'; \varepsilon_c) Z_3(q', 0; \varepsilon_c)$$

$$+ \frac{2}{\pi^2} \int q'^2 dq' \Delta Z_2(p, q'; 0, q'; \varepsilon_c) \chi^{-1}(q'; \varepsilon_c)$$

$$\int dq'' Z_3(q', q''; \varepsilon_c) \tau^{-1}(q''; \varepsilon_c) T(q'', 0; \varepsilon_c), \tag{6.70}$$

where

$$\Delta Z_1(p, 0; 0, 0; \varepsilon_c)$$
$$= Z_1(p, 0; \varepsilon_c) - Z_1(0, 0; \varepsilon_c); \ \Delta Z_1(p, q; 0, q; \varepsilon_c) = Z_1(p, q; \varepsilon_c) - Z_1(0, q; \varepsilon_c)$$
$$\text{and } \Delta Z_2(p, q; 0, q; \varepsilon_c) = Z_2(p, q; \varepsilon_c) - Z_2(0, q; \varepsilon_c) \tag{6.71}$$

The resulting integral Eq. (6.70), as demonstrated in the work of Afnan and Phillips [40] has the kernel that goes to zero faster as $q' \to \infty$ than the kernel of integral Eq. (6.68) does and is therefore expected to admit a unique solution to Eq. (6.70).

Further, since off-the-energy shell scattering amplitude $T(p, q; E)$ is known to be symmetric in the variables p and q, we write Eq. (6.67) for $T(q, p; E)$ by interchanging $p \Leftrightarrow q$ when $k = 0$ and $E = \varepsilon_c$ to get

$$T(q, p; \varepsilon_c) = Z_1(q, p; \varepsilon_c) + \frac{1}{\pi} \int dq' Z_1(q, q'; \varepsilon_c) \tau^{-1}(q'; \varepsilon_c) T(q', p; \varepsilon_c)$$
$$+ \frac{2}{\pi} \int q'^2 dq' Z_2(q, q'; \varepsilon_c) \chi^{-1}(q'; \varepsilon_c) Z_3(q', p; \varepsilon_c)$$
$$+ \frac{2}{\pi^2} \int q'^2 dq' Z_2(q, q'; \varepsilon_c) \chi^{-1}(q'; \varepsilon_c)$$
$$\int dq'' Z_3(q', q''; \varepsilon_c) \tau^{-1}(q''; \varepsilon_c) T(q'', p; \varepsilon_c) \qquad (6.72)$$

and the half-off-shell amplitude when $q = k = 0$ and $E = \varepsilon_c$ is

$$T(0, p; \varepsilon_c) = Z_1(0, p; \varepsilon_c)$$
$$+ \frac{1}{\pi} \int dq' Z_1(0, q'; \varepsilon_c) \tau^{-1}(q'; \varepsilon_c) T(q', p; \varepsilon_c)$$
$$+ \frac{2}{\pi} \int q'^2 dq' Z_2(0, q'; \varepsilon_c) \chi^{-1}(q'; \varepsilon_c) Z_3(q', p; \varepsilon_c)$$
$$+ \frac{2}{\pi^2} \int q'^2 dq' Z_2(0, q'; \varepsilon_c) \chi^{-1}(q'; \varepsilon_c)$$
$$\int dq'' Z_3(q', q''; \varepsilon_o) \tau^{-1}(q''; \varepsilon_c) T(q'', p; \varepsilon_c) \qquad (6.73)$$

Subtracting Eq. (6.73) from Eq. (6.72), we have

$$T(q, p; \varepsilon_c) = T(0, p; \varepsilon_c) + \Delta Z_1(q, p; 0, p; \varepsilon_c) + \frac{1}{\pi} \int dq' \Delta Z_1$$
$$(q, q'; 0, q'; \varepsilon_c) \tau^{-1}(q'; \varepsilon_c) T(q', p; \varepsilon_c) + \frac{2}{\pi} \int q'^2 dq' \Delta Z_2$$
$$(q, q'; 0, q'; \varepsilon_c) \chi^{-1}(q'; \varepsilon_c) Z_3(q', p; \varepsilon_c) + \frac{2}{\pi^2} \int q'^2 dq' \Delta Z_2(q, q'; 0, q'; \varepsilon_c) \chi^{-1}(q'; \varepsilon_c)$$
$$\int dq'' Z_3(q', q''; \varepsilon_c) \tau^{-1}(q''; \varepsilon_c) T(q'', p; \varepsilon_c) \qquad (6.74)$$

On comparing the kernels of Eq. (6.74) with those appearing in Eq. (6.70), we note that they have essentially the same structure. We shall now check numerically that the amplitude $T(q, p; \varepsilon_c)$ satisfying this integral equation, which, by construction, is real and symmetric, would give a unique numerical solution, independent of the cutoff value Λ.

To write the equation for the scattering amplitude at energy E, we start from Eq. (6.67) and introduce a function $X(p, q; E)$ collecting the inhomogeneous terms and rewrite the equation as:

$$T(p, q; E) = X(p, q; E) + \frac{1}{\pi} \int dq' \frac{q'^2}{q'^2 - k^2 - i\varepsilon} Z_1(p, q'; E)\tau^{-1}(q'; E)T(q', q; E)$$

$$+ \frac{2}{\pi^2} \int q'^2 dq' Z_2(p, q'; E)\chi^{-1}(q'; E)$$

$$\int dq'' \frac{q''^2}{q''^2 - k^2 - i\varepsilon} Z_3(q', q''; E)\tau^{-1}(q''; E)T(q'', q; E), \qquad (6.75)$$

where

$$X(p, q; E) \equiv Z_1(p, q; E) + \frac{2}{\pi} \int dq' q'^2 Z_2(p, q'; E)\chi^{-1}(q'; E)Z_3(q', q; E)$$
$$(6.76)$$

Now, we subtract from Eq. (6.75) the equation corresponding to the amplitude $T(p, q; \varepsilon_c)$ when $E = \varepsilon_c$ to get

$$T(p, q; E) = T(p, q; \varepsilon_c) + [X(p, q; E) - X(p, q; \varepsilon_c)]$$

$$+ \frac{1}{\pi} \int dq' \{ \frac{q'^2}{q'^2 - k^2 - i\varepsilon} Z_1(p, q'; E)\tau^{-1}(q'; E)T(q', q; E) \\ -Z_1(p, q'; \varepsilon_c)\tau^{-1}(q'; \varepsilon_c)T(q', q; \varepsilon_c)\}$$

$$+ \frac{2}{\pi^2} \{ \int q'^2 dq' Z_2(p, q'; E)\chi^{-1}(q'; E)$$

$$\int dq'' \frac{q''^2}{q''^2 - k^2 - i\varepsilon} Z_3(q', q''; E)\tau^{-1}(q''; E)T(q'', q; E)$$

$$- \int dq' q'^2 Z_2(p, q'; \varepsilon_c)\chi^{-1}(q'; \varepsilon_c)$$

$$\int dq'' Z_3(q', q''; \varepsilon_c)\tau^{-1}(q''; \varepsilon_c)T(q'', q; \varepsilon_c)\} \qquad (6.77)$$

To write this equation for half-off-shell scattering amplitude, we put $q = k$ where $E = k^2/2m_d + \varepsilon_c$ and express the scattering amplitude as $T_k(p, E)$. Note that the two amplitudes $T_k(p, E)$ and $T_k(p, \varepsilon_c)$ are different. However, for these functions appearing within the integrals, we make a simplifying assumption, viz. $T_k(q', E) = T_k(q', \varepsilon_c)$ and $T_k(q'', E) = T_k(q'', \varepsilon_c)$. This approximation is based on the fact that below the breakup threshold, the values of k, for which we are interested here, are going to be small as compared to ε_c. With this simplification, Eq. (6.77) reduces to:

$$T_k(p, E) = T_k(p, \varepsilon_c) + [X_k(p, E) - X_k(p, \varepsilon_c)]$$

$$+ \frac{1}{\pi} \int dq' \{ \frac{q'^2}{q'^2 - k^2 - i\varepsilon} Z_1(p, q'; E) \tau^{-1}(q'; E) \\ - Z_1(p, q'; \varepsilon_c) \tau^{-1}(q'; \varepsilon_c) \} T_k(q', E)$$

$$+ \frac{2}{\pi^2} \{ \int q'^2 dq' Z_2(p, q'; E) \chi^{-1}(q'; E)$$

$$\int dq'' \frac{q''^2}{q''^2 - k^2 - i\varepsilon} Z_3(q', q''; E) \tau^{-1}(q''; E)$$

$$- \int dq' q'^2 Z_2(p, q'; \varepsilon_c) \chi^{-1}(q'; \varepsilon_c)$$

$$\int dq'' Z_3(q', q''; \varepsilon_c) \tau^{-1}(q''; \varepsilon_c) \} T_k(q'', E) \tag{6.78}$$

This is the integral equation which is subject to calculating the half-off-shell amplitude for different values of k below the breakup threshold for $n - ^{19}C$ scattering. To reduce it to the on-shell amplitude, we put

$$T_k(p, E)|_{p=k} = -\tau(0, \varepsilon_c) \exp(i\delta) \sin \delta / k = -\tau(0, \varepsilon_c) \frac{1}{k \cot \delta - ik} = -\tau(0, \varepsilon_c) f_k,$$

from where we calculate the s-phase shifts δ for different k, viz.

$$\text{Re}[f_k]^{-1} = k \cot \delta, \text{ where } f_k = \tau^{-1}(0 \cdot \varepsilon_c) T_k(p, E)|_{p=k}$$

$$\text{and } \tau^{-1}(0, \varepsilon_c) = -2 \left(\frac{d}{a} \right) \varepsilon_c.$$

Turning over to discuss the computational details for the expressions of scattering amplitude thus obtained, we consider, as a first step, the integral Eq. (6.74), to check numerically whether this equation is independent of the cutoff value of the parameter Λ. Using the values for the two-body ($n - ^{18}C$) binding energy $\varepsilon_c = 140 \text{ keV}$, the $n-n$ scattering length $a_{nn} = -23.69 \text{ fm}$ and the three-body scattering length, $A_s = 12 \text{ fm}$, as input parameters, we have computed the amplitude by inverting a matrix of size (80×80) for the ($q \times p$) variables, by choosing different values of the cutoff parameter Λ lying between 30 and 60. The analysis shows that the resulting computed matrix $T(p, q; \varepsilon_c)$ is indeed Λ independent. We then proceed to compute the integral Eq. (6.78) for the half-off-shell scattering amplitude, $T_k(p; E)$ for $k \neq 0$. Here, we notice that we again encounter the singularity in the two-body propagator for $k \neq 0$ and therefore employ the same prescription suggested in [38] by rotating the integration variables $q' \to q'_1 \exp(-i\phi)$ and $q'' \to q''_1 \exp(-i\phi)$ thereby deforming the contour from its original position along the real axis to the position rotated by the angle ϕ. Clearly, the angle ϕ is merely a parameter to be chosen in such a way that the new contour lies as far away as possible from the singularities. By this operation, it is ensured that the kernel of the integral equation is analytically continued so as to become compact. These values are restricted by the requirement that the resulting scattering amplitude calculated from the integral equation must be unitary; i.e., it

Table 6.12 Comparing the values of k with $\mathrm{Im}\left(f_k^{-1}\right)$ at a few typical values of ϕ of the Off-shell scattering amplitude $T(p, k; \varepsilon_c)$ (Eq. (6.74))

k in units of α	Φ (radians)	$\mathrm{Re}(f_k^{-1})$	$\mathrm{Im}(f_k^{-1})$	p in units of α
0.04928	0.7972	−0.07958	−0.04827	0.4927
0.5698	0.6297	0.09404	−0.5698	0.5698
0.06487	0.4149	−0.0944	−0.0648	0.0645

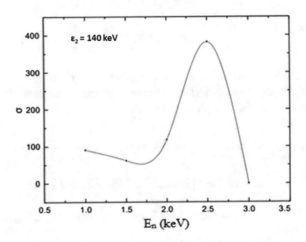

Fig. 6.24 A plot of total scattering cross section versus E_n (keV)

must satisfy the condition $\mathrm{Im}(f_k^{-1}) = -k$. Since this requirement of unitarity is very much crucial in the context of effective field theory (see, e.g., Ref. [93]), we have checked that it should be satisfied at every stage of the calculations. In the table given below, we show comparison of $\mathrm{Im}(f_k^{-1})$ with a few typical values of k (Table 6.12).

Figure 6.24 depicts a plot of total scattering cross section σ (in mb) versus incoming neutron kinetic energy (keV) below the breakup threshold for $n-{}^{19}\mathrm{C}$ scattering for the case when ${}^{19}\mathrm{C}$ is bound of $n-{}^{18}\mathrm{C}$ system with binding energy 140 keV. It is heartening to note that the present calculations carried out in the framework of effective field theory reproduce a remarkable qualitative agreement with the resonant structure already predicted in the potential approach.

While there appears a qualitative trend showing a resonant structure in the region of kinetic energy between 2 and 3 keV (centered around 2.5 keV), but the peak here does not appear to be as sharp. It may be quite possible that the simplifying approximation, viz. $T_k(q'; E) \approx T_k(q'; \alpha_c)$ just does not hold good in the region of n-core binding energy equal to or greater than 140 keV. These considerations, therefore, lead us to undertake a detailed numerical investigation keeping in view the following aspects: (i) to compute the scattering amplitude without resort to the approximation stated above; (ii) to carry out the numerical analysis with double precision, particularly because the renormalization procedure involves too many subtractions at various stages introducing thereby the possible truncation error; (iii) to check how far the lower cutoff value of Λ (which is admissible for stable results) has an effect on the scattering amplitude; and (iv) to study the effect of variation of the scattering length parameters from the ground state energy to the regime of excited state energies. The last point is expected to shed light on the correlation between three-body scattering lengths and the corresponding binding energies for both ground state and excited states on the lines similar to the recent attempt [113] on the study of weakly bound three-body systems to estimate the degree of universality with short-range interactions. The study of $n-{}^{19}C$ scattering on these aspects will be the subject for future publication.

Chapter 7
Summary and Future Outlook

This brief report describes a systematic development of theoretical attempts carried out during the last more than two decades to investigate the structural properties of 2n-rich halo nuclei employing three-body approach using two-body separable potentials. In the first two chapters, we lay down the groundwork to present the essentials of two-particle scattering theory and highlight the salient features of Faddeev's three-body scattering formalism. There we also present a comparative study to show how the basic difficulties of disconnectedness of the kernel and boundary conditions encountered in the standard Lippmann Schwinger equation for three-body scattering problem are tackled in a natural way using three-body Schrodinger equation with separable potentials.

Chapter 3 is devoted to the study of Efimov effect wherein the basic steps to obtain the expression for attractive long-range effective potential for three identical bosons using the proper boundary conditions for the resonant pair interactions are outlined. An important consequence of this attractive long-range three-body potential, $1/R^2$ as shown by Efimov, is that as long as the two-body scattering length $|a|$ is sufficiently large compared to the range r_0, there is a sequence of three-body bound states whose binding energies form a geometric series lying in the interval between $\hbar^2/mr_0^2$ and $\hbar^2/ma^2$. In fact, Efimov effect describes a universal low-energy behavior of the three-particle system. It is for this reason of the universality character that the effect can be realized and observed in various physical contexts, e.g., in atomic, molecular, nuclear or condensed matter physics. In the context of separable potential approach, this singular behavior of $1/R^2$ type is translated into momentum space as a singularity of $1/p$ type in the limiting case as two-body scattering length $|a| \to \infty$.

Another important area which has attracted in recent years a great deal of attention in the domain of low-energy few body dynamics is the effective field theory. This topic has been briefly touched in Chap. 4. The general principle of effective field theory is illustrated by considering two-nucleon system, where the scattering length $|a| \gg |r_{\text{eff}}| \approx r_0$ and the contact or pionless effective field theory can be used. As we

V. S. Bhasin and I. Mazumdar, *Few Body Dynamics, Efimov Effect and Halo Nuclei*, SpringerBriefs in Physics, https://doi.org/10.1007/978-3-030-56171-0_7

include the higher-order loop diagram, the ultraviolet divergence arising from high momentum modes of loop diagram in the Lagrangian can be regulated by a floating ultraviolet cutoff Λ. As all the observables are to be independent of the cutoff, the theory is normalized by absorbing the Λ-dependence into the effective field theory couplings, thereby making the theory to be model independent.

To extend this theory to three-body sector, three-body integral equations for bound and scattering states using the two-body zero-range or contact interactions were obtained long ago [36]. Recently, Bedaque and collaborators [22] derived this equation within the framework of effective field theory. In order to renormalize this equation, they introduced a three-body force in the leading order three-body EFT equation and adjusted this force to reproduce the experimental $1 + 2$ scattering length. Following an alternative approach used in ref. [40] where the three-body scattering length parameter was introduced to renormalize the amplitude of the EFT equation, we employed a similar procedure to set up integral equations starting from separable potential approach to obtain the EFT equations for studying the resonant structure in $n - {}^{19}C$ scattering, where ${}^{19}C$ is assumed to consist of dimer $(n - {}^{18}C)$ system. The detailed renormalization procedure along with numerical calculations is deferred to Chap. 6.

Chapter 5 highlights the experimental status reviewing the attempts made in the early 1990s on the discovery of halo structure in light neutron and proton drip-line nuclei. In particular, in the case of 2n-halo nuclei experimental attempts made and the evidence produced for the existence of three-body structure with two neutrons orbiting around the core, specifically in nuclei like, ${}^{11}Li$ has been highlighted.

Finally, in Chap. 6, theoretical investigations carried out during the last more than two decades on studying the structural properties of halo nuclei within the framework of three-body formalism have been reviewed. Primarily, the ground state properties (binding energy, matter radius, n-n and n-core correlations) and resonant states of ${}^{11}Li$, probability distributions in ${}^{11}Li$ and β–decay of ${}^{11}Li$ into halo analog states and to ${}^{9}Li$ +deuteron channel have been reported in the first part of the chapter. The second part is devoted to the search for Efimov states in halo nuclei like ${}^{14}Be$, ${}^{19}B$, ${}^{22}C$ and ${}^{20}C$. The promising candidate, ${}^{20}C$, which is characterized by a $(2n-{}^{18}C)$ three-body system and where one of the neutrons halo is weakly bound to the core ${}^{18}C$ has been investigated in detail, where the possibility of the occurrence of more than one Efimov state is predicted. A detailed analysis of the movement of the Efimov states leads to the demonstration of resonances in $n - {}^{19}C$ scattering near threshold. The appearance of resonances in elastic scattering cross section, as the Efimov states just become unbound, led us to tie this result to yet another interesting result that a general resonance profile is asymmetric, which was pointed out by Fano [103] over four decades ago. While such asymmetric profiles have been widely observed in atoms and molecules, resonances in nuclear and particle physics generally show symmetric Lorentzian or Breit–Wigner shapes. It is the combination of the features of the Efimov and Fano phenomena which, in fact, holds promise for the possible observation of such resonances with the asymmetry being used as a diagnostic for the Efimov effect.

We also made an attempt to enlarge the scope of present analysis for studying the Efimov effect beyond ^{20}C and found that heavy nuclei such as ^{32}Ne and ^{38}Mg, exhibit a structure dynamically similar to ^{20}C [108]. In fact, we generalized the analysis by including a system of a very heavy core ($A = 100$) with two halo neutrons, to find the conditions for the occurrence of the Efimov states in these systems [109] and see how these states evolve into the Feshbach-type resonances.

7.1 Future Outlook

The key question to be addressed in the context of Efimov–Fano resonances predicted in ^{20}C, in the present analysis, is: How far the Efimov criterion, viz. $|a|/r_{eff} \gg 1$, is satisfied? Experimentally, our knowledge regarding the two-body $n - {}^{18}$C interaction, employed in ^{20}C, is rather limited; the only parameter experimentally known is the binding energy. Obviously, this is not sufficient to determine precisely the parameters of the realistic two-body potential. The question raised above is also important from the point of view of the scaling criterion, i.e., $E_T^{(n+1)}/E_T^{(n)} \rightarrow \exp(-2\pi/s_0) \rightarrow 1/515.03$, as $n \rightarrow \infty$ at $a = \pm\infty$. Efimov has also shown that this scaling limit also holds for large scattering length. In the present *case*, this ratio in the case of first excited state to the ground state is predicted to be around 1/20, which is quite far off from the scaling law. Interestingly, experimental studies with ultra-cold atoms also reported only a couple of Efimov states without exhibiting the universal scaling. In atomic systems, as, for instance, in the case of Cs atoms, their interaction measurements in the lowest internal state are experimentally determined by applying varying magnetic fields, thereby changing the s-wave scattering length all the way from $-2500\ a_0$ to $+1600\ a_0$, where a_0 is Bohr's radius. When an Efimov state interacts with the continuum threshold at negative scattering lengths a, three free atoms in the ultra-cold limit resonantly couple to a trimer—evolving into a tri-atomic Efimov resonance. Another type of Efimov resonance appears at positive values of scattering length, a, for collisions between a free atom and a dimer. In the domain of nuclear physics with halo nuclei, we do not enjoy such a privilege to get the scattering lengths changed by applying varying magnetic fields. However, by changing the two-body binding energy in case of ^{19}B and ^{22}C halo nuclei, we found [94] that excited Efimov states could appear by getting the scattering lengths at -179.6 fm for $n - {}^{17}$B system at 0.67 keV and at -121.5 fm for $n - {}^{20}$C system at 1.46 keV. In the case of positive scattering length, we have the example of ^{20}C where the occurrence of Efimov states has been studied in detail. These objects may, therefore, be considered, from the experimental standpoint, as possible candidates to search for the occurrence of Efimov resonances. In particular, we should expect that experimentally, in addition to the scattering of low-energy $n - {}^{19}$C scattering, the photo-excitation processes, through the absorption of gamma rays from the ground states of these halo targets near threshold energies, should possibly show Efimov–Fano resonances. Let us hope that with the advent of new machines for generating and studying neutron-rich halo nuclei, we can look forward to the observation of the effect predicted by Efimov long ago.

A good amount of interest is also being currently witnessed in some attempts [113], which explore the possibility for the occurrence of 'super Efimov effect' in condensed matter physics, thereby opening new vistas for universal few body physics. The authors employ anisotropic Heisenberg model to show that collective excitations inducing two magnon resonances can result in the binding of three magnons, the binding energy spectrum obeying the 'doubly exponential scaling,' i.e., $E_n \approx \exp(-2\exp(-3\pi n/4))$. For this reason, it should, more appropriately, be termed as a '**second-generation**' Efimov effect.' The authors in [113] also propose several approaches to experimentally realize such an effect in quantum magnets. Extending theoretically the 'normal' Efimov effect, which by itself, has only been experimentally established in certain idealized situations, to the so-called second-generation Efimov effect and looking for its experimental realization appears, at this stage, to be a far-fetched possibility.

Bibliography

1. L.D. Faddeev, in *Three Body Problem in Nuclear and Particle Physics*, ed. by J.S.C. McKee, P.M. Rolph (North Holland, Amsterdam, 1970).
2. C. Lovelace, in *Strong Interactions and High Energy Physics*, ed. by R.G. Moorhouse (Oliver and Boyd, London, 1964)
3. A.N. Mitra, in *Advances in Nuclear Physics*, ed. by M. Baranger, E. Vogt, vol II (Plenum Press, Inc., New York, 1969)
4. R.D. Amado, Ann. Rev. Nucl. Sci. **19**, 635 (1969)
5. L.M. Delves, A.C. Phillips, Rev. Mod. Phys. **41**, 497 (1969)
6. W. Sandhas, Acta Phys. Austr. (Suppl. IX), 57 (1972)
7. L.D. Faddeev, Scattering theory of three particle system, Sov. Phys. JETP **12** (1961); See also Mathematical aspects of the three body problem in the Quantum Scattering Theory, Israel Program for Scientific Translations, Jerusalem (1965).
8. C. Lovelace, Phys. Rev. **135**, B1225 (1964)
9. A.N. Mitra, Nucl. Phys. **32**, 529 (1962)
10. A.N. Mitra, V.S. Bhasin, Phys. Rev. **131**, 1265 (1963)
11. A.G. Sitenko, V.F. Kharchenko, Nucl. Phys. **49**, 15 (1963)
12. R.D. Amado, Phys. Rev. **132**, 485 (1963)
13. V. Efimov, Phys. Lett. **33B**, 563 (1970). See also, Sov. J. Nucl. Phys. **12**, 589 (1971); Nucl. Phys. **A210**, 157 (1973); Comments Nucl. Part. Phys. **19**, 271 (1990); Phys. Rev. C **47**, 1876 (1993). See also, R.D. Amado, J.V. Noble, Phys. Rev. **D5**, 1992 (1972)
14. See, e.g., Efimov physics: a review, by P. Naridon, S. Endo, Reports on Prog. In Physics **80**(5), 056001
15. Tanihata et.al., Phys. Lett. **B160**, 380 (1985), ibid. **B206**, 592 (1988); See also T. Kobayashi et al., Phys. Rev. Lett. **60**, 2599 (1988); T. Kolayashi et al., Phys. Lett. **B232**, 51 (1989)
16. T. Hoshino, H. Sagawa, A. Arima, Nucl. Phys. **B506**, 271 (1990); G. Bertsch, J. Foxwell, Phys. Rev. C **41**,1300 (1990)
17. R. Arnold, Phys. Lett. B **197**, 398 (1987); ibid. **281B**,16 (1992)
18. T. Kraemer et al., Nature **440**, 315 (2006)
19. F. Ferlaino et al., Phys. Rev. Lett. **103**, 043201 (2009), ibid., **101**, 023201 (2008); See also S. Knoop et al., Nat. Phys. **5**, 227 (2009)
20. E. Braaten, H.W. Hammer, Phys. Rev. Lett. **87**, 160407–1 (2006)
21. N.P. Mehta et al., Phys. Rev. Lett. **103**, 153201 (2009). See also Y. Yang, B.D. Esry, Phys. Rev. Lett. **102**, 133201 (2009)
22. E. Braaten, H.W. Hammer, Phys. Rep. **428**, 259 (2006)
23. For more details, see, eg., E.W. Schmid, H. Ziegelmann, in *The Quantum Mechanical Three Body Problem in Vieweg Tracts in Pure and Applied Physics*, vol 2 (1074)

© The Author(s), under exclusive license to Springer Nature Switzerland AG 2021
V. S. Bhasin and I. Mazumdar, *Few Body Dynamics, Efimov Effect and Halo Nuclei*,
SpringerBriefs in Physics,
https://doi.org/10.1007/978-3-030-56171-0

24. D. Fedorov, A. Jensen, Phys. Rev. Lett. **71**, 4103 (1993)
25. V. Efimov, Sov. J. Nucl. Phys. **29**, 546 (1979)
26. V. Efimov, Nucl. Phys. A **362**, 45 (1981)
27. E.G. Tracheko, Phys. Lett. **98**, 328 (1981)
28. V.S. Bhasin, V.K. Gupta, Phys. Rev. C **25**, 1675 (1982)
29. R.D. Amado, J.V. Noble, Phys. Lett. **35B**, 25 (1971)
30. S.K. Adhikari, L. Tomio, Phys. Rev. **C26**, 83 (1982); ibid., **C26**, 77 (1082)
31. V.S. Bhasin, W. Sandhas, in *Few Body Problems in Physics*, vol II, p. 365, ed. by B.Zeitnitz (Elsevier Science Publishers B.V., 1984)
32. G.P. Lepage, in arXiv:nucl-th/9706029
33. See e.g., Lectures on the theory of the nucleus by A.G. Sitenko, V.K. Tartakoviskii, translared by P.J. Shephered (Pergamon Press, 1975)
34. U.van Kolck, Prog. Part. Nucl. Phys. **43**, 337 (1999)
35. D.B. Kaplan, M.J. Savage, M.B. Wise, Phys. Lett. B **424**, 390 (1998); See also Nucl. Phys. B **534**, 329 (1998) and Phys. Rev. C **59**, 617 (1999)
36. G.V. Skorniakov, K.A. Ter-Martirosian, Sov. Phys. JETP **4**, 648 (1957)
37. G.S. Danilov, Sov. Phys. JETP **13**, 349 (1961)
38. V.V. Komarov, A.M. Popova, Nucl. Phys. **54**, 278 (1964)
39. S.K. Adhikari, T. Frederico, I.D. Goldman, Phys. Rev. Lett. **74**, 487 (1995)
40. I.R. Afnan, D. Phillips, Phys. Rev. C **69**, 034010 (2004)
41. I. Tanihata et al., Phys. Rev. Lett. **55**, 2676 (1985)
42. B. Blank et al., Z. Phys. A **343**, 375 (1992)
43. G. Hansen, B. Jonson, Euro. Phys. Lett. **4**, 409 (1987)
44. T. Kobayashi et al., Phys. Rev. Lett. **60**, 2560 (1988)
45. M.H. Smedberg et al., Phys. Lett. B **452**, 1 (1999)
46. T. Minamisono et al., Phys. Rev. Lett. **69**, 2058 (1992)
47. R. Kanungo et al., Phys. Lett. B **571**, 21 (2003)
48. P. Egelhof et al., Euro. Phys. J A **15**, 27 (2002)
49. M. Takechi et al., Euro. Phys. J A **25**, 217 (2007)
50. I. Tanihata, H. Savajols, R. Kanungo, Prog. In Part. Nucl. Phys. **68**, 215 (2013)
51. K. Hagino, N. Takahashi, H. Sagawa, Phys. Rev. C **77**, 054317 (2008)
52. K. Hagino, H. Sagawa, P. Schuck, J. Phys. G **37**, 064040 (2010)
53. K. Hagino, H. Sagawa, Phys. Rev. C **89**, 014331 (2014)
54. D.S. Ahn et al., Phys. Rev. Lett. **123**, 212501 (2019)
55. K.J. Cook et al., Phys. Rev. Lett. **124**, 212503 (2020)
56. T.B. Webb et al., Phys. Rev. Lett. **122**, 122501 (2019)
57. Y. Kondo et al., Phys. Rev. Lett. **116**, 102503 (2016)
58. S. Leblond et al., Phys. Rev. Lett. **121**, 262502 (2018)
59. Y. Blumenfeld, T. Nilsson, P. Van Duppen, Phys. Scr. **T152**, 014023 (2013)
60. O. Wieland et al., Phys. Rev. Lett. **102** (2009)
61. Halo Nuclei, J.S. Khalili, IOP Science (2017)
62. R. Chatterjee, R. Shyam, Prog. in Part. Nucl. Phys. **103**, 67 (2018)
63. T. Nakamura, H. Sakurai, H. Watanabe, Prog. In Part. Nucl. Phys. **97**, 53 (2017)
64. G. Bertsch, J. Foxwell, Phys. Rev. **C41**, 1300 (1990)
65. G. Bertsch, H. Esbensen, A. Sutich, Phys. Rev. C **42**, 758 (1990)
66. S. Dasgupta, I. Mazumdar, V.S. Bhasin, Phys. Rev. **50**, R550 (1994)
67. N.A. Orr et al., Phys. Rev. Lett. **69**, 2050 (1992)
68. T. Kobayashi et al., in *Proceedings of Second International Conference on Radioactive Nuclear Beams*, ed. by T. Delbar (Louvain-la-Neuve, Belgium, 1991)
69. K. Ieki et al., Phys. Rev. Lett. **70**, 730 (1993)
70. W.R. Gibbs, A.C. Hayes, Phys. Rev. Lett. **67**, 1395 (1991)
71. D.F. Fedorov, A.S. Jensen, K. Riisager, Phys. Lett. B **312**, 1 (1993)
72. S. Kumar, V.S. Bhasin, Phys. Rev. C **65**, 034007 (2002)

73. S. Kumar, V.S. Bhasin, Modern Phys. Lett. **A547** (2004); See also PRAMANA-J. Phys. **63**, 509 (2004)

74. I. Mukha et al., Nucl. Phys. A **616**, 201c (1997)

75. M.J.G. Borge et al., Nucl. Phys. A **613**, 199 (1997)

76. M.V. Zhukov et al., Phys. Rev. C **52**, 2461 (1995)

77. M. Zinser et al., Phys. Rev. Lett. **75**, 1719 (1995)

78. B.M. Young et al, Phys.Rev. **C49**, 279 (1994)

79. E. Garrido, D.V. Fedorov, A.S. Jensen, Phys. A **700**, 117 (2002)

80. F. Humbert et al., Phys. Lett. B **347**, 198 (1995)

81. P.G. Hansen, Nucl. Phys. A **553**, 89c (1993)

82. T. Myo, K. Kato, S. Aoyama, K. Ikeda, Phys. Rev. C **63**, 054313 (2001)

83. D.R. Harrington, Phys. Rev. B **69**, 139 (1965)

84. G. Cattapan et al., Nucl. Phys. A **241**, 204 (1975)

85. Mukha et al., Phys. Lett. B **367**, 65 (1996)

86. K. Rissager, Nucl. Phys. A **616**, 169c (1997)

87. D. Baye, P. Descouvemont, Nucl. Phys. A **481**, 445 (1988)

88. Y. Yamaguchi, Phys. Rev. **95**, 1628 (1954)

89. T. Teranishi et al., Phys. Lett. B **407**, 110 (1997)

90. Mazumdar, V.S. Bhasin, Phys. Rev. **56**, R5 (1997)

91. G. Audi, A.H. Wapstra, Nucl. Phys. A **565**, 66 (1993)

92. M. Thoennessen et al., Phys. Rev. C **63**, 014308 (2000)

93. P.G. Hansen, A.S. Jensen, B. Jonson, Ann. Rev. Nucl. Part. Sci. **45**, 591 (1995); V. Efimov, Nucl. Phys. A **216**, 157 (1973)

94. I. Mazumdar, V. Arora, V.S. Bhasin, Phys. Rev. C **61**, 051303(R) (2000)

95. A.E.A. Amorim, T. Frederico, L. Tomio, Phys. Rev. C **56**, R378 (1997)

96. R.D. Amado, J.V. Noble, Phys. Rev. D **5**, 1992 (1972)

97. S.K. Adhikari, L. Tomio, Phys.Rev. C **26**, 83 (1982): S.K. Adhikari, A. C. Fonseca, L. Tomio, ibid. **26**, 77 (1982)

98. T. Nakamura et al., Phys. Rev. Lett. **83**, 1112 (1999)

99. V. Arora, I. Mazumdar, V.S. Bhasin, Phys. Rev. C **69**, 061301 (R) (2004)

100. M.T. Yamashita, T. Frederico, A. Delfino, L. Tomio, Phys. Rev. **A66**, 052702 (2002); also see V. Efimov, Sov. J. Nucl. Phys. **29**, 546 (1979)

101. E. Baslev, J.M. Combes, Commun. Math. Phys. **22**, 280 (1971); Y. Matsui, Phys. Rev. C **22**, 2591 (1980)

102. I. Mazumdar, A.R.P. Rau, V.S. Bhasin, Phys. Rev. Lett. **97**, 062503 (2006)

103. U. Fano, Phys. Rev. **124**, 1866 (1961)

104. T. Kraemer et al., Nature (London) **440**, 315 (2006); B.D. Esry, C.H. Greene, ibid. **440**, 289 (2006); C. Day, Phys. Today **59**(4) 18 (2006)

105. See, for instance, U. Fano, J.W. Cooper, Rev. Mod. Phys. **40**, 441 (1968); U. Fano, A.R.P. Rau, *Atomic Collisions and Spectra* (Academic, Orlando, 1986)

106. See, for instance, S. Bar-Ad et al., Phys. Rev. Lett. **78**, 1363 (1997) D. Neuhauser, T.J. Park, J.I. Zink, ibid., **85**, 5304 (2000), V. Madhavan et al., Science **280**, 567 (1998); K. Kobayashi et al., Phys. Rev. Lett. **88**, 256806 (2002); J. Kim et al., ibid **90**, 166403 (2003)

107. [107] The similar $2p^2$ state with the same overall ^1S symmetry will also be strongly mixed in but, for our discussion here, we ignore such complications. See C.D. Lin. Adv. At. Mol. Phys. **22**, 77 (1986)

108. G. Audi, A.H. Wasptra, Nucl. Phys. A **729**, 737 (2003)

109. I. Mazumdar, V.S. Bhasin, A.R.P. Rau, Phys. Lett. B **704**, 51 (2011)

110. P.F. Bedaque, H.W. Hammer and U.van Kolck, Phys. Rev. Lett. 82,463(999); Nucl. Phys.A 676, 357(2000).

111. B. Blankenleider, J. Gagelia, aXiv:nucl-th/00900701 (2000)

112. H.W. Hammer, T. Mehen, Nucl. Phys. A **690**, 535 (2000)

113. Y. Nishida, S. Moroz, D.T. Son, Phys. Rev. Lett. **110** 235301 (2013); See also Y. Nishida, Y. Kato, C.D. Batista, Nature Phys. **8**, 93 (2013)